# ELECTROMECHANICAL PRINCIPLES OF WIND TURBINES

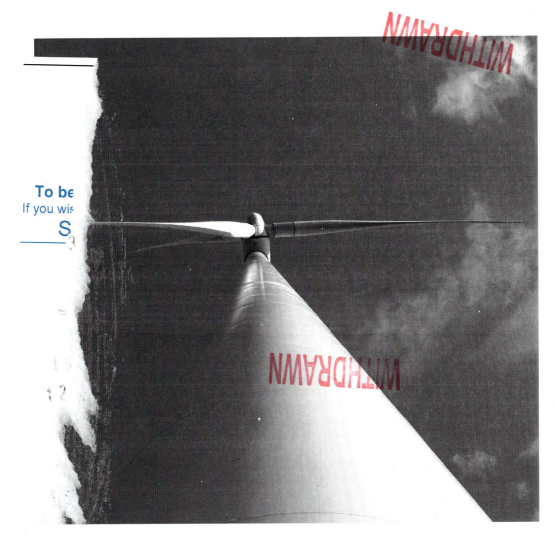

## Keith Plantier
## & Karen Mitchell Smith

TSTC
Publishing

ISBN 978-1-934302-54-5 (softback)
ISBN 978-1-934302-79-8 (ebook)

TSTC Publishing
Texas State Technical College Waco
3801 Campus Drive
Waco, TX 76705

http://publishing.tstc.edu/

Publisher: Mark Long
Project Manager: Grace Arsiaga
Printing production: Bill Evridge
Illustrations: Kai Jones, Eric Larson, Jessica Luckey, George Miller, Charles Ragona & Bethany Stewart
Interior design and layout: Stephen Tiano (steve@tianobookdesign.com)
Indexer: Michelle B. Graye (indexing@yahoo.com)
Copyeditor: Katharine O'Moore-Klopf, ELS (editor@kokedit.com)

Manufactured in the United States of America

First edition

# Table of Contents

# Introduction

Wind turbines are relatively simple machines. The wind blows, turning the blades attached to the hub. This rotational energy is transferred from the hub to the shaft and then into a gearbox. The gearbox merely takes the slow rotational energy and torque of the turbine and converts it to a high-speed rotation that is fed into the generator. From there, the generator produces the necessary electricity, which is transmitted to the substation for consumption. This is really an oversimplification of the process, as there are thousands of components—both stationary and moving—that are required for all of this to happen continuously. That turbines are located in remote areas increases the difficulty of maintaining them. In general, wind farms are located where people and infrastructure (roads) are not. Therefore, industry is forced to set up shop in a location central to the mass acreage. Getting repair parts and deliveries becomes a major issue. For example, the wind turbine technician might have a turbine 50 miles away that has a fault. An hour's drive, a 20-minute climb, and the technician starts troubleshooting, only to find out that repair requires a part not available on-site or one that is located back at the remote-control area.

*TSTC Wind Turbine—Roscoe, TX*

In the early 1980s, the first modern wind turbine was installed in California. By the end of 2008, wind-powered generators were providing approximately 30 gigawatts (GW) of electricity in the United States. Although wind energy provides only about 1% of electricity worldwide, it is a rapidly growing resource, increasing the amount of available electricity by more than 500% globally between 2000 and 2007. Wind farm construction has increased at a steady rate since 2008, resulting in a need for wind turbine technicians and their supervisors to have a strong knowledge in the area of electrical and mechanical components.

Although wind turbines are the most prevalent pieces of equipment associated with a wind park, they are not the only equipment in such parks. Other pieces of infrastructure, such as service roads, an underground distribution system, an overhead collection system, substations, meteorological equipment, and operation and maintenance facilities, along with wind turbines, comprise a wind farm.

The most obvious structure on a wind farm is of course the wind turbine itself. It is composed of the tubular steel tower structure; the nacelle, which is the big box on top of the tower structure; and the rotor, which is composed of the central hub and blades. Inside the nacelle are some of the major components in a wind turbine. These components include the main shaft, the main shaft bearing, the bed plate (which is the supporting structure for all of the major components found inside the nacelle), the gearbox, the generator, and control devices.

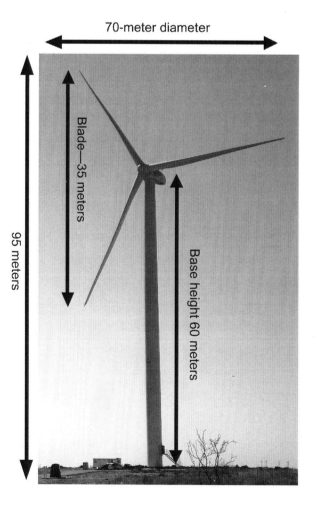

# Evolution of U.S. Commercial Wind Technology

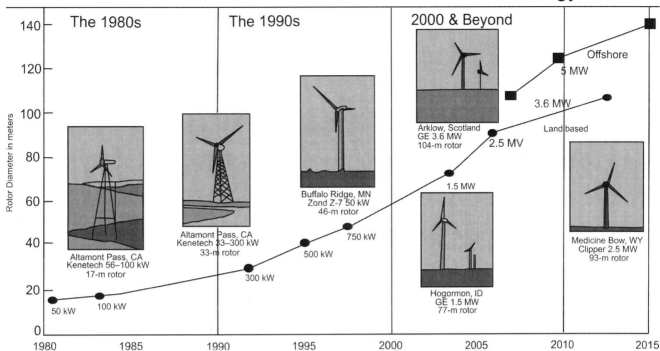

As a wind turbine technician, you will need to be familiar with all of the equipment inside a turbine. The job is autonomous by nature, and there will not be a lot of help readily available while you are at a remote work site. It is critical that you be able to perform normal maintenance tasks, identify issues that may arise during your time spent at the turbine, and accurately troubleshoot when problems come up. To accomplish this task, you should have a very thorough knowledge of the turbine. The chapters in this book will help you become familiar with the materials and electromechanical equipment associated with wind turbines as well as with some problems that can occur during design, manufacture, transportation, installation, and operation of the turbine. After all, technicians can help alleviate some of the problems in the supply chain because they are the end users of the equipment and experience a lot of the problems firsthand.

This book covers topics associated with wind turbine materials and electromechanical equipment associated with wind turbines. The learning objectives that you will be able to meet after reading this book are as follows:

1.  Describe the effects of heat generation on various materials and heat-control mechanisms; define the effects of machining and heat treating on metals as they relate to predictable failures

2.  Identify gel coats, ultraviolet (UV) characteristics, flexibility, and impact resistance of various coatings and how they are applied

3.  Identify types and specifications of fasteners; recognize the effects of torque, lubricants, and hydraulic bolt stretchers, tensioners, and high torque on fasteners

4.  Inspect gears, scaling, types of gearboxes (hybrid, planetary versus helical or parallel shaft) and probable causes of failure

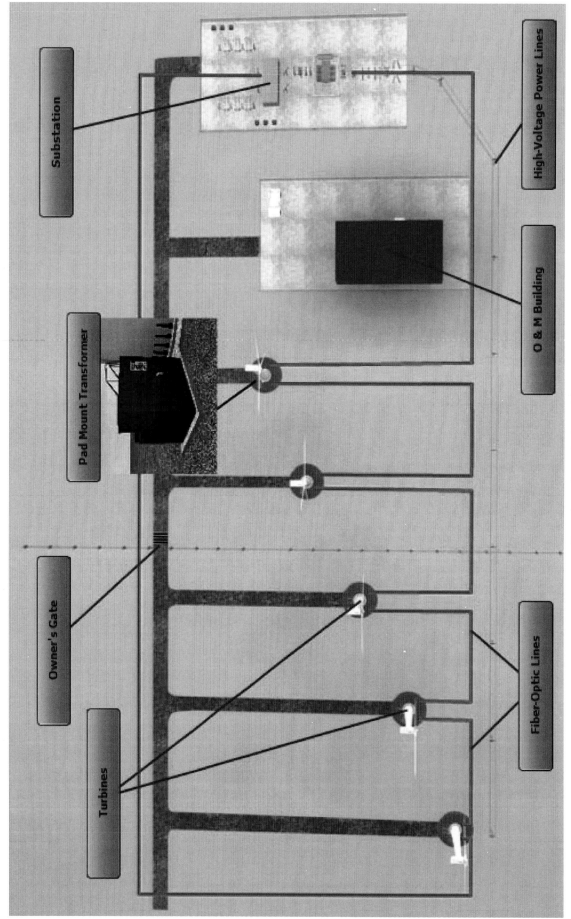

*Overview of a typical wind farm*

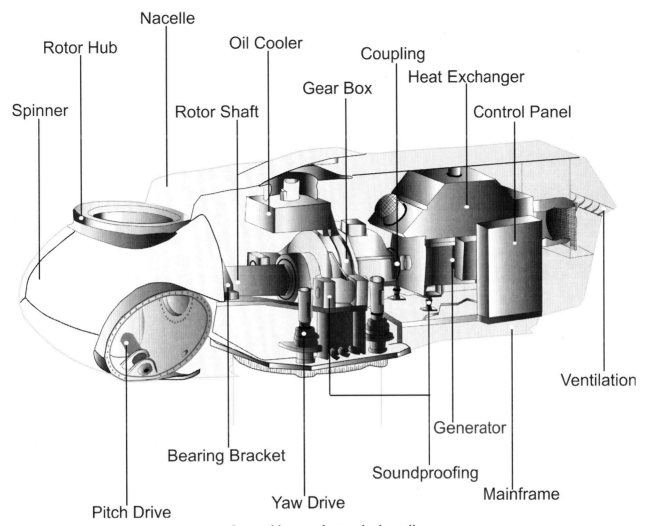

Rotor Hub

Nacelle

Oil Cooler

Coupling

Heat Exchanger

Spinner

Rotor Shaft

Gear Box

Control Panel

Ventilation

Generator

Bearing Bracket

Soundproofing

Mainframe

Pitch Drive

Yaw Drive

*Internal layout of a standard nacelle*

5. Identify revolutions per minute (rpm), gear ratios, and failure mechanisms

6. Identify type, application, and compatibility of different lubricants

7. Identify electrical control system components such as circuit-protection devices, sensors, relays, contactors, actuators, timers, counters, motors, and various types of DC and AC drives

8. Convert units between metric and U.S. standard

9. Demonstrate safety procedures required by OSHA (Occupational Safety and Health Administration) 1910, the NFPA (National Fire Protection Association), IEEE (formerly Institute of Electrical and Electronics Engineers, Inc.) 519, International Electric Code, and National Electric Code standards

Here is a brief description of what is covered in each chapter:

- **Chapter 1:** An introduction to wind turbines that covers basic unit conversions. The chapter discusses the units—and their standard conversions—that a technician is expected to be familiar with in the field.

- **Chapter 2:** An introduction to safety related to wind turbines and the wind industry in general. This chapter briefly addresses safety procedures required by OSHA 1910, NFPA, IEEE 519, International Electric Code and National Electric Code standards and how they are related to the wind industry.

- **Chapter 3:** A discussion of fasteners and their applications in the wind industry. Topics covered include identification of types and specifications of fasteners; the effects of torque, lubricants, hydraulic bolt stretchers, tensioners, and high torque on fasteners; identification, application, and compatibility of different lubricants; and the proper use of torque wrenches.

- **Chapter 4:** A thorough discussion of gears, gearboxes, and gear trains. Topics covered include gear inspection, scaling, types of gearboxes (hybrid, planetary versus helical or parallel shaft), and probable causes of failure. Also covered are rpm, gear ratios, and failure mechanisms, along with types applications and compatibility of different lubricants.

- **Chapter 5:** A brief discussion of electrical control systems and their components. to the text describes and identifies electrical-control-system components such as circuit-protection devices, relays, contactors, actuators, timers, counters, motors, and various types of AC and DC drives, along with the purpose of each.

- **Chapter 6:** An explanation of materials associated with wind turbines and description of the effects of heat generation on various materials and heat-control mechanisms. Wind turbines come in many sizes and configurations and are built from a wide range of materials. Accordingly, this chapter covers gel coats and their functions, UV characteristics, flexibility and impact resistance of various coatings and how they are applied, and effects of machining and heat treating on metals as they relate to predictable component failures.

The topics covered in this book will allow you, as a technician, to become familiar with the components of a typical wind turbine and how they relate to and interact with one another. Additionally, you will learn about aspects of the operation and maintenance of a turbine and how you to successfully operate and maintain turbines safely, reliably, and efficiently.

## Chapter One
# Unit Conversions

## Learning Objectives

*As a wind turbine repair technician, you will be faced with equipment and formulas that involve a variety of measurement units. You will need to be able to convert units quickly from one form to another, with the units depending on the situation at hand.*

*In this chapter, you will learn:*

- *Definitions of various units of measure*
- *Addition, subtraction, multiplication, and division of units associated with numbers*
- *Conversion of one set of units in a function or equation into another, using equivalencies set for mass, length, area, volume, time, energy, and force*
- *Application of the concepts of dimensional consistency to determine the units of any term in a function*

## *Introduction*

Dimensions are our basic concepts of measurement. Length, time, mass, temperature, and volume are all examples of dimensions. To measure dimensions and express a given amount, we must have units. Some examples of units are pounds, feet, centimeters, liters, and seconds. If you have a certain amount of material that you would like to sell to someone else, you both must speak the same "unit language" to agree on not only the amount of material you want to sell but also the amount of money you want to sell the material for.

In June 1215, the Magna Carta addressed this very issue: "There shall be one measure of wine throughout our kingdom, and one measure of ale, and one measure of grain ... and one breadth of cloth. ... And of weights it shall be as of measures." The standards set forth in this historic document were not substantially revised until the nineteenth century. Even when the American colonies declared independence from England, the standards of weights and measures remained the same, as the fledgling country continued to do business with England.

However, no such uniform standard was used throughout the rest of Europe. Weights and measures differed from country to country and even from town to town within a country. In May 1790 during the French Revolution, to combat this discrepancy, the National Assembly of France enacted a decree that called on the French Academy of Sciences to join with the Royal Society of London to "deduce an invariable standard for all of the measures and all weights." Because the English already had a firm system of measurement units in place, they declined to participate, and the French developed what is known as the metric system. As scientists began to observe new phenomena and experiment with new medicines, the scientific community showed a preference for the metric system for three reasons:

1.  The metric system was intended to be an international system.
2.  The system was theoretically independently reproducible.
3.  The system's decimal nature is simple to work with.

These scientists began to derive new units that were based on the laws of physics for newly observed physical phenomena, working from the known units of mass, weight, and length in the metric system. Eventually the British and Americans adapted their common systems to the new technology in the area of commerce too, in spite of the fact that other countries were embracing the metric system.

Because of problems in the specification of units for electricity and magnetism, numerous international conferences were called. Finally, in 1960, during the eleventh General Conference on Weights and Measures, the Système International (SI) classification of units was settled on. To this day, the United States is the last major country not using the metric system.

# Terms

In the field, you will encounter a variety of terms that may not be familiar to you. Learning the terms used in the SI will help you be more successful. Often the derived units have special names that honor physicists. For example, a Newton (N) is the measurement of a unit of force. The equation is $kg \ m \ s^{-2}$.

Here are some of the other SI units you will encounter:

## Meter
Symbolized by the lower-case m, the meter measures linear distance. The meter is defined as the distance a beam of light travels through a perfect vacuum in 1/299,792,458, or $3.33564095 \ 10^{-6}$, of a second. You can visualize a meter as the approximate distance an adult covers in a single brisk stride. It is slightly longer than 36 in or an English yard.

Units larger or smaller than meters are defined as follows:

*   Millimeter (mm): 0.001
*   Micrometer (μm): $10^{-6}$
*   Nanometer (nm): $10^{-9}$
*   Kilometer (km): 1000 m

## Kilogram

The kilogram is expressed as kg and is the base unit of SI mass. It is important to understand that mass is not the same thing as weight. Weight changes with gravitational force, but mass does not. The prototype for a kilogram is a sample of platinum-iridium alloy that is kept at the International Bureau of Weights and Measures in Sèvres, France. That prototype will have the same mass of 1 kg no matter where it is located, on the moon, in space, or on Earth. On Earth, however, it weighs 2.2 pounds (lb). Without gravity, it would weigh substantially less.

Units that define mass larger or smaller than 1 kg are

- Gram (g): 0.001 kg
- Milligram (mg): $10^{-6}$ kg
- Microgram (μg): $10^{-9}$ kg
- Nanogram (ng): $10^{-12}$ g
- Metric ton: 1000 kg

## Second

The SI measures time in seconds, abbreviated as the letter s. Besides being 1/60 of a minute, a second is the time it takes for a beam of light to travel $2.99792458 \times 10^{8}$ m through space. The moon is a little more than 1 light-second away from Earth.

Units of time smaller than a second are

- Millisecond (ms): 0.001 s
- Microsecond (μs): $10^{-6}$ s
- Nanosecond (ns): $10^{-9}$ s

## Joule

The standard unit of energy in the SI system is expressed as a joule, abbreviated as an uppercase letter J. A joule is the equivalent of a newton-meter (N·m) and can be expressed as *unit mass × unit distance squared per unit time squared*, or $1J = 1 \text{ kg} \cdot \text{m}^2/\text{s}^2$.

## Watt

Abbreviated by an uppercase letter W, the watt is the standard unit of power. One watt is equivalent to 1 joule expended per second, or 1 J/s. Power measures the rate at which energy is produced, radiated, or consumed.

## Coulomb

The *coulomb* is the standard unit for electric charge quantity and is abbreviated as an uppercase letter C. This is the amount of electric charge found in $6.241506 \times 10^{18}$ electrons.

## Volt

The volt is the standard unit of electrical potential and is abbreviated as an uppercase letter V. One volt is equal to 1 joule per coulomb, or 1 J/C. Most car batteries in the United States have between 12 and 13.5 V.

### Ohm

An ohm is the standard unit of electrical resistance and is expressed with the Greek letter for omega, which is $\Omega$. The ohm is equivalent to 1 volt per ampere, or 1 V/A.

### Ampere

The ampere is the standard unit of measure for electric current and is abbreviated as an uppercase letter A. A flow of approximately $6.241506 \times 10^{18}$ electrons per second, past a given focal point in an electrical conductor, is 1 ampere, or 1 A.

### Hertz

The standard unit of frequency is hertz, abbreviated as Hz. The hertz is used to express audio or wireless frequencies and measures frequency cycles per second.

## Why Unit Conversions Are Important in Your Job

You may be wondering why a wind turbine repair technician would need to know unit conversions. Even though you will most likely be working on turbines only in the United States, those turbines will have parts and equipment manufactured in European countries. Sometimes, even equipment manufactured in the United States may employ the metric system of measurement. For example, TECO-Westinghouse Manufacturing Company makes pressure gauges for measuring the hydraulic pressure of lubrication systems. Rather than using pounds per square inch (psi), these gauges use bars as a measurement. As a technician, you will need to know how to convert bars to pounds per square inch.

| Conversion of Bars to Pounds per Square Inch | | | | |
|---|---|---|---|---|
| Pascal (Pa) | bar (bar) | atmosphere (atm) | torr (Torr) | lbf/in² (psi) |
| **1 Pa** | 1.00E-05 | 9.8692E-05 | 7.5006E-04 | 1.4504E-04 |
| **1 bar** 100,000 | | 0.98692 | 750.06 | 14.50377 |
| **1 atm** 101,235 | 1.01325 | | 760.00 | 14.696 |
| **1 torr** 133.322 | 1.3332E-03 | 1.3158E-03 | | 1.9337E-03 |
| **1 psi** 6,894.76 | 6.8948E-02 | 6.8046E-02 | 51.72 | |

## Converting Dimensions and Units

When you convert dimensions, is it important to remember that you can perform mathematical functions such as adding, subtracting, or equating only with units that measure like dimensions of an object. Just as an apple cannot be added to an orange to become an orange, a kilogram cannot be added to a kilowatt; a kilogram is a measure of mass, whereas a kilowatt is a measure of power. A pound, however, can be added to a gram, once the necessary conversion has been performed.

For multiplying and dividing, you can work with unlike units—for example, *20 meters ÷ 5 seconds = 4 meters/second*—but you cannot cancel out unlike units. Units contain a large amount of information that cannot be ignored. In the example earlier in this paragraph, we learned that a given object moves at the velocity of 4 meters every second. If we had ignored one of those units, we would have lost vital information.

---

**EXAMPLE 1.1. PROBLEMS OF DIMENSIONS AND UNITS**

Add the following equations:

a. 4 miles + 15 seconds =

b. 3 horsepower + 100 watts =

*Solutions*

4 miles + 15 seconds

a. This problem is unworkable because you are being asked to add two unlike dimensions. Miles measure distance, and seconds measure time.

3 horsepower + 100 watts

b. In this problem, we are dealing with two like dimensions—both measure power. The units, however, are unlike. To solve the problem, convert both units to either horsepower (hp) or watts (W). If you refer to the conversion tables in the front of the book or appendices, you will see that 1 hp = 0.7457 kilowatts (kW). One kw is equal to 1000 W (the root kilo- means "1000"). Therefore, multiply 0.7457 kW by 1000 to convert horsepower to watts. Thus:

1 hp = 0.7457 kW     *or*     745.7 W

Solve the problem by substituting 745.7 W for 1 hp in the original equation; this will produce equivalent units that can then be added together:

3(745.7 W) + 100 W = 2217.1 W + 100 W = 2317.1 W

---

You see from table in this chapter for converting bars to pounds per square inch that most physical or chemical values can be expressed equivalently with a variety of units, as long as the units are alike. Conversion factors must be constructed that can be multiplied by or divided into the original unit to derive the equivalent unit. Conversion factors are ratios that describe the relationship of one unit to another.

A conversion factor has two essential properties. First it must be a mathematically alternate expression for the value 1. This allows either multiplication or division of an expression by the conversion factor without altering the value of the expression. The second property is that it must contain units of both the original and the desired equivalent expression.

Conversion factors are constructed from identity equations that define the units involved—for example, the identity relationship that defines a centimeter (cm) in terms of a meter (m):

$$100 \text{ cm} = 1 \text{ m}$$

can be used to construct the conversion factor:

$$100 \, \frac{\text{cm}}{\text{m}}$$

This conversion factor describes the relationship between centimeters and meters. It meets the two criteria of

1. Being an alternate expression of the value 1
2. Containing both units of the conversion process (cm and m)

Therefore it can be used to rewrite any original expression involving meters into an equivalent expression involving centimeters.

As already mentioned, unit conversion results from multiplying or dividing the original expression by the conversion factor to derive the equivalent expression in the alternate unit. The decision to multiply or divide by the conversion factor is made to ensure cancellation of all original units leaving only the desired alternate units. Although this book offers handy conversion factor tables inside the front cover as well as in the appendices, it is not always possible in the field to have access to such information. For this reason, technicians must be familiar with the mathematical equation for converting units.

An expression written as meters (for example, 10 m) can be converted to an equivalent expression written as centimeters:

$$10 \text{ m} \left( \frac{100 \text{ cm}}{\text{m}} \right) = (10)(100) \text{ cm} = 1000 \text{ cm}$$

The original expression, 10 m, is multiplied by the conversion factor 100 cm/m to allow the original unit (m) to be canceled, leaving the desired alternate unit (cm).

Reversing the process can convert an expression written as centimeters (for example, 1000 cm) to an equivalent expression written in meters. This time, the original expression is divided by the conversion factor 100 cm/m (note that this is the same as multiplying by the reciprocal conversion factor of 1 m/100 cm to allow cancellation of the unit centimeter).

$$1000 \text{ cm} \left(\frac{\text{m}}{1000 \text{ cm}}\right) = \left(\frac{1000}{100}\right) \text{m} = 100 \text{ m}$$

The process of unit conversions often involves the use of several conversion factors that provide intermediate sets of units. This technique is convenient when a single conversion factor is unknown and is required when more than one unit must be converted. For example, an expression written in inches (such as 10 in) can be converted to meters using a single (although obscure) conversion factor:

$$10 \text{ in} \left(\frac{0.0254 \text{ m}}{\text{in}}\right) = 0.254 \text{ m}$$

or by a combination of two (more commonly known) conversion factors:

$$10 \text{ in} \left(\frac{2.54 \text{ cm}}{\text{in}}\right)\left(\frac{\text{m}}{100 \text{ cm}}\right) = 0.254 \text{ m}$$

When more than one unit is to be converted, then more than one conversion factor must be used. For example speed expressed in the English units of miles per hour (for example, 10 mph), can be converted to metric units of meters per second (m/s) as follows:

$$\frac{10 \text{ mi}}{\text{h}}\left(\frac{1 \text{ h}}{60 \text{ min}}\right)\left(\frac{1 \text{ min}}{60 \text{ s}}\right)\left(\frac{5280 \text{ ft}}{\text{mi}}\right)\left(\frac{12 \text{ in}}{\text{ft}}\right)\left(\frac{2.54 \text{ cm}}{\text{in}}\right)\left(\frac{\text{m}}{\text{cm}}\right) = \left(\frac{4.47 \text{ m}}{\text{s}}\right)$$

**Using Conversion Factor Tables**
In preceding example, we converted miles per hour to meters per second using familiar unit conversion factors as if we did not have a conversion table to look at. Now we will look at a conversion problem similar to that one; however, be sure to reference the conversion factor tables found inside the front cover of this book.

**EXAMPLE 1.2. CONVERTING UNFAMILIAR RATIOS**

A wind turbine with a radius of 77 m turning at approximately 22 rpm (revolutions per minute) has the potential to produce 1.5 to 2 megawatts (MW) of electrical power. At 22 rpm, the tip of the blade is spinning at 177.4 m/s. Use the conversion factors from the tables inside the front cover of this book to convert this tip speed to miles per hour.

*Solution*

$$\frac{177.4 \cancel{m}}{\cancel{s}} \left| \frac{6.214 \cdot 10^{-4} \text{ mi}}{\cancel{m}} \right| \left\| \frac{3600 \cancel{s}}{1 \text{ h}} \right\| = 10.28 \, \frac{\text{mi}}{\text{h}}$$

## Power of Wind Turbine (Based on Swept Area of Rotor)

Now we will stretch your new skills a little further. A typical GE 1.5 MW wind turbine has a blade length of 253 feet (ft). Air density (R) is the mass per unit volume of the Earth's atmosphere. Air density decreases with increasing altitude, which is similar in principle to air pressure. It is also inversely proportional to temperature. At sea level and 20°C (standard temperature and pressure), air has a density of approximately 1.225 kg/m$^3$ (kilograms per cubic meter).

If air density is $4.426 \times 10^{-5}$ lbf/in$^3$ (pound force per cubic inch) and wind speed is 20 mph, using the information provided, determine the power produced by the wind turbine. Determine power of the wind turbine if the speed is increased by 10% (in miles per hour). Is the change significant or insignificant? Explain.

Given:

$$1 \text{ kg} = 2.205 \text{ lbf}$$
$$1 \text{ m}^3 = 35.315 \text{ ft}^3 \text{ (cubic feet)}$$
$$1 \text{ ft}^3 = 1728 \text{ in}^3 \text{ (cubic inches)}$$
$$1 \text{ mile} = 1.609 \text{ km (kilometers)}$$

$$P_{WT} = \frac{1}{2} \times \rho \times A_{\text{swept area}} \times V^3 \times C_p \times N_g \times N_b$$

Where:

$P_{WT}$ = Power of the wind turbine in watts

$R$ = air density in kg/m$^3$

$A_{\text{swept area}}$ = rotor swept area, exposed to the wind in square meters (m$^3$)

$V$ = wind speed in m/s

$C_p$ = Coefficient of performance (assume 0.35)

$N_g$ = Generator efficiency (assume 0.80)

$N_b$ = Gearbox/bearings efficiency (assume 0.90)

**PROBLEM 1**
Determine the swept area of the rotor in square feet $(\text{ft}^2)$.

**PROBLEM 2**
Convert the swept area from square feet to square meters.

**PROBLEM 3**
Convert 20 mph to meters per second.

**PROBLEM 4**
Determine speed in miles per hour when increased by 10%. Convert to meters per second.

$$\left(\frac{20 \text{ mi}}{\text{h}}\right) + \left(\frac{20 \text{ mi}}{\text{h}}\right) \times (10\%)$$

$$= \left(\frac{20 \text{ mi}}{\text{h}}\right) + \left(\frac{2 \text{ mi}}{\text{h}}\right)$$

$$= 22 \, \text{mi}\big/\text{h}$$

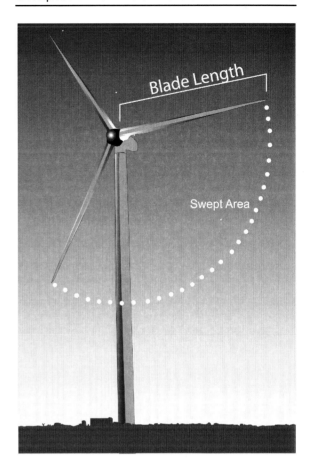

**PROBLEM 5**
Convert $4.426 \times 10^{-5}$ lbf/in³ to kilograms per cubic meter.

**PROBLEM 6**
Determine power for the wind turbine at 10 mph and the final power after a 10% increase in wind speed.

## Conclusion

In repairing wind turbines, you will encounter equipment and machined parts whose sizes are measured in lots of different units. To ensure that all parts of a turbine will work together despite their being measured in disparate units, you will have to be able to quickly convert units to the proper forms. This chapter taught you the definitions of many units of measure, how to work mathematically with quantities expressed in various units of measure, and how to convert a set of units to another set using standard equivalencies.

## *References and Additional Resources*

### Information on the Metric System
http://www.mathleague.com/help/metric.htm

http://www.visionlearning.com/library/module_viewer.php?mid=47

http://www.unc.edu/~rowlett/units/metric.html

### Free Online Unit Conversion Software
http://www.unitconversion.org/

### Tutorials for Unit Conversion
http://www.oakroadsystems.com/math/convert.htm

http://www.allaboutcircuits.com/vol_5/chpt_1/10.html

## *Review Questions*

1. Why are all units of measure not the same worldwide?

2. How was the metric system decided on?

3. What country is the only one in the world that does not use the SI units of measure?

4. What are the two essential properties of a conversion factor?

5. Define the following units of measure:

   • Joule

   • Bar

   • Coulomb

   • Pounds per square inch

   • Meter

   • Kilowatt

   • Hertz

6. Name at least two areas on the job as a wind turbine technician for which you will need to be able to convert units of measure.

7. In the conversion of units of measure, which mathematical functions can be performed on like units and which ones can be performed on unlike units?

8. Complete this sentence: Conversion factors are ratios that _____.

9. Even though conversion charts exist, such as the ones on the inside front cover of this book, why do technicians need to be familiar with the mathematical equation for conversion of units of measure?

10. How did most metric units get their names?


# Application Exercises

1. Convert the following:

   a. 0.75 kg to milligrams

   b. 1500 mm to kilometers

   c. 2390 g to kilograms

   d. 50 mph to kilometers per second

   e. 33°C to degrees Fahrenheit

2. Use the following table for conversions a through j:

   1 kilometer = 0.6 miles

   1 mile = 1.6 km

   1 mile = 5280 feet

   1 fathom = 6 feet

   1 meter = 3.3 feet

   1 kilogram = 2.2 pounds

   $C = \frac{5}{9} \times (F - 32)$

   1 m = 100 cm

   a. A dolphin dives to 24 ft. How many fathoms is this below the surface of the water?

   b. The deepest part of the ocean is the Mariana Trench; its depth is 11.03 km. How many feet is that?

   c. The largest whale measured 33.27 m in length. How many feet was that?

   d. The average temperature of the oceans is 3.9°C. What is this in degrees Fahrenheit?

   e. The highest tides on planet Earth occur near Wolfville, in Nova Scotia's Minas Basin. The water level at high tide can be as much as 16 m higher than at low tide! How many feet is this?

    f. Sunlight only reaches 30 to 120 m under the waves. Convert this to feet.

    g. A blue whale can weigh up to 280,000 lb. That is larger than the largest dinosaur. How many kilograms is that? (Round your answer to the nearest whole kilogram.)

    h. Studies suggest that the expected global warming from the greenhouse effect could raise sea level approximately 100 cm in the next century or two. How much is that in feet?

    i. If all the ice in glaciers and ice sheets melted, the sea level would rise by about 80 m, or _____ ft.

    j. The longest river in the world is the Nile River, in Egypt, Africa. It is 4,160 mi long and flows northward into the Mediterranean Sea. How long is it in kilometers?

3. Conduct research on the Internet concerning the Mars Climate Orbiter. Discuss in your own words what happened and the results of the incident. Additionally, discuss how the circumstances of this event could affect you in your daily work environment in the field (any industry).

## Chapter Two

# Safety on the Job

## Learning Objectives

*Every career comes with its own set of safety regulations and potential hazards. The job of wind turbine repair technician is no different in that respect from any other job, so you must know the specific safety issues that pertain to turbine repair and maintenance.*

*In this chapter, you will learn about:*

- *The origins and purposes of the Occupational Safety and Health Administration (OSHA)*
- *Personal safety responsibility*
- *Personal protective equipment and its uses*
- *The two-person rule*
- *The need for first aid and cardiopulmonary resuscitation*
- *OSHA 1910 regulations and training courses: OSHA 10- and 30-hour training*
- *Crane safety*
- *Extreme-weather safety*
- *Whistle-blower techniques*

## *Introduction*

The Occupational Safety and Health Administration (OSHA) reported in 2002 that every year, approximately 6,000 Americans die from work-related injuries and another 50,000 die from illnesses caused by exposure to workplace hazards. Additionally, 6 million workers sustain nonfatal workplace injuries, at an annual cost to U.S. businesses of more than $125 billion (*Taking Charge: Your Education, Your Career, Your Life*, Karen Smith, 2008). Since 1971, OSHA has been overseeing workplace safety and providing guidance to employers and employees alike.

When you start your career in wind energy, you will attend specific on-the-job safety training meetings and be required to meet various goals and standards. Details may vary depending on which company you work for and how the company trains its employees, but having familiarity now with the potential hazards and regulations that you will face will better prepare you for a safe and successful career as a wind turbine technician.

*Trained worker performing inspections outside on a DeWind D8.2 Wind Turbine Generator.*

# Job 1

You have probably heard the saying "Safety is job 1." This statement is true in any industry but is especially relevant in regarding wind turbines. On a wind farm, significant risk factors contribute to the potential for severe personal injury to repair and maintenance technicians, other personnel on-site, landowners, subcontractors, livestock, and members of the general public. Hazards may also be present that can cause significant damage or loss of property, structures, equipment, and vehicles.

As a maintenance or repair technician, your job responsibilities may include site and equipment inspections, preventative maintenance, emergency repairs, oversight of subcontractors, site-access management, ground control, structure or site security, driving, and parts transportation. Because your job title is likely to entail several of these activities, plus managerial tasks, it is crucial that you come away from this chapter with a perspective that is much broader than safety alone and is focused on the concept of *total risk management.*

Total risk management attempts to take into consideration controls for all risk factors involved in the production of energy by wind turbines. The potential for loss or accidents includes but is not limited to personal injury, damage to or destruction of structures on-site, production equipment, vehicles and mobile equipment, and transmission and distribution lines and can be caused by a variety of hazards, such as fire, lightning, electrical surge, flood, theft and vandalism, and aircraft. Loss-prevention methods available to help limit these exposures are extensive and are typically developed by your company, but it is your responsibility to ensure that these controls are in place. A constant part of your normal daily routine will be to identify situations or conditions that can create loss, communicate these to appropriate supervisory personnel, warn others around you, and help implement control measures. Unless your personal safety is at stake, never leave an unsafe situation or condition without eliminating it.

Members of the public may be on-site from time to time; this could include inspectors, subcontractors, power transmission personnel, landowners, and other visitors. It may be your responsibility to escort these people or supervise activities for them. The potential for injury to members of the public who access the site creates significant liability exposure for your employer. Also, you may be found personally liable if you fail to correct known unsafe conditions, situations, or activities that result in injury to others. Your role in site security may also be vital in preventing unauthorized persons from accessing the site, by your maintaining security systems, ensuring that entry and gate locking systems are in place, and so on.

Your personal safety and the safety of fellow employees will be a vital focal point of any activity you perform. OSHA requires that your employer have specific safety rules, written policies, and written procedures developed for each task that you perform on the job. Your company is required to provide you with appropriate training and equipment for each job activity, the risks and hazards involved with the activity, and the controls required to be in place to safely perform the activity. In fact, your daily on-the-job activities will most likely begin with a mandatory safety strategy meeting.

On the job, you will be at risk for physical injuries such as slips, trips, cuts, falls, blows, contusions, crushes, burns, and contact with energy sources, as well as occupational health injuries resulting from exposure to chemicals, noise, heat, cold, vibration, and repetitive stress. You could be exposed to wild animals indigenous to your work area (snakes, mountain lions, bears, skunks, badgers, bobcats, raccoons), livestock, and insects such as spiders, wasps, and bees. Some of the generic controls and training programs that will help you stay safe on the job are outlined in this chapter. When reviewing these, keep in mind that your employer will develop policies, procedures, and training programs that are specific to your particular work environment and equipment, and even if you come out of college with a strong base in safety knowledge, your employer will require you to participate in on-the-job training, both for your career training and for safety objectives.

## *Personal Protective Equipment*

In an article titled "Project Developer Takes Safety Lessons to New Heights," written by Steve O'Gorman and published in the journal *North American Wind Power* in 2005, wind technician Ian Lindsell said, "We go up almost daily to the nacelle—the unit at the top of the tower that houses the equipment—to handle annual preventative or routine maintenance. Most of the mechanical stuff is high up in the nacelle—the generators, gearbox, main shaft, hub, and blades. Eighty percent of our time is spent there and twenty percent is spent down tower."

Many of the potential hazards you will face as a wind turbine technician are the same as those faced by any utility worker. Risk of electrical shock, mechanical injury, lifting injury, fire, repetitive motion strain, and noise-related injury are all factors you may deal with. However, wind turbine technicians deal with one hazard uncommon to most utility workers: You will be required to climb towers that are up to 260 feet tall. This, in effect, is like climbing the face of a 20-story building. Additionally, wind turbine maintenance is often conducted in extremely confined spaces. Maintenance technicians should expect to wear full-body harnesses all day. Proper use of personal protective equipment (PPE) could make the difference between life and death in a crisis situation or in some of the unsafe conditions described in this chapter's introduction. Commonly employed PPE in the wind energy industry includes

- Full-body harnesses
- Self-rescue and personal evacuation equipment
- Hard hats
- Steel-toed boots
- Climbing gloves and high-voltage-rated gloves
- Safety glasses
- Ear protection

## *Two-Person Rule*

Because of the dangers inherent with working 260 feet off the ground, almost all wind energy companies employ the two-person rule, and some require more than two people to work together. If you were inside the turbine and were to experience a severe injury or life-threatening situation such as an electric shock or heart attack, you would need a partner to help you reach safety. Areas such as the tower legs and inside the blades are confined spaces that present especially difficult challenges for extracting injured or ill workers. Additionally, the danger of life-threatening falls is always present.

Dave West, president of and instructor for Vertical Systems International, a company that provides training in mountain safety and technical rescue, rope access, and fall protection, believes that wind turbine repair presents specific challenges that must be proactively addressed for repair technicians to confidently and successfully complete their tasks. Owing to their environments, he said, "these towers can be slippery during wet weather. In addition to this, tower climbers experience very high wind and extreme fatigue from climbing up 300-foot ladders while wearing heavy work clothing, tools, and equipment. For these reasons, training has to be thorough and detailed." According to West, in the event of a fall, it is imperative that a worker is rescued in a timely manner. Suspension trauma, also known as harness hang syndrome, restricts blood flow in a situation when a person falls and hangs immobile for too long, allowing blood to pool in the lower extremities. It can be a life-threatening condition. Symptoms of harness hang syndrome appear within 7 to 20 minutes and begin with an feeling of illness, something like having the flu, followed by excessive sweating, nausea, dizziness, and hot flashes. Brain function deteriorates rapidly. If the situation continues, the patient will have difficulty breathing, experiencing an elevated heart rate, cardiac arrhythmias, and an abrupt increase in blood pressure, followed by unconsciousness and death. The two-person rule helps to prevent this potentially fatal situation, as coworkers are trained and available to lower the fallen worker to the ground before suspension trauma sets in. As a precaution, each set of turbine technicians carries a bag of rescue equipment, such as the one shown in this chapter, into the tower with them.

## *First Aid and Cardiopulmonary Resuscitation*

Because tower rescue can be so difficult, proper employee training in basic first aid and cardiopulmonary resuscitation (CPR) procedures can often make the difference between life and death in a situation in which stopgap measures must be used before rescue can occur. So important is this training that OSHA regulations require each two-person team have at least one CPR-certified member. However, many companies

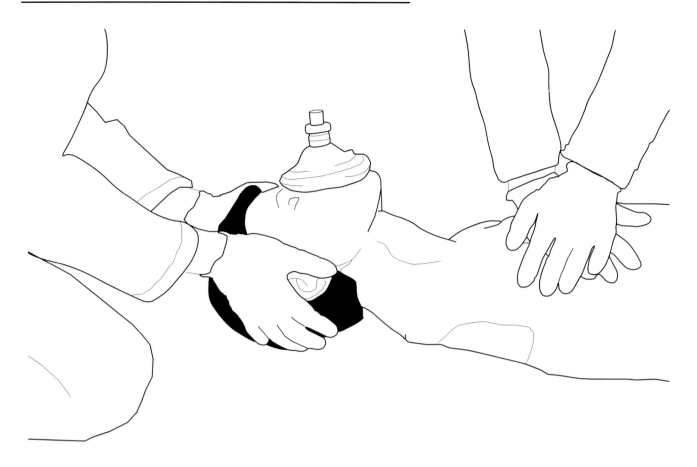

require CPR certification or the ability to become certified as a job qualification for all employees. As a result, in many colleges and training programs, you will learn first aid and CPR as part of your curriculum.

## OSHA 1910 Regulations

Because of the relative newness of wind power as an energy utility, OSHA has not created any regulations specific to the industry. Instead, wind turbine technicians receive training in OSHA 1910 regulations during OSHA 10-hour and 20-hour training sessions.

OSHA 1910 regulations (http://www.osha.org) address general industry safety concerns. For example, OSHA 1910.21 deals with handling safety issues on walkways, stairs, and other work surfaces. A section specific to wind turbines is 1910.27, which deals with fixed ladders, such as those inside turbine towers:

- **1910.27(c)(4):** "Clearance in back of ladder. The distance from the centerline of rungs, cleats, or steps to the nearest permanent object in back of the ladder shall be not less than 7 inches, except that when unavoidable obstructions are encountered, minimum clearances as shown in figure D-3 shall be provided."

- **1910.27(d)(2)(ii):** "All landing platforms shall be equipped with standard railings and toeboards, so arranged as to give safe access to the ladder. Platforms shall be not less than 24 inches in width and 30 inches in length."

- **1910.27(d)(2)(iii):** "One rung of any section of ladder shall be located at the level of the landing laterally served by the ladder. Where access to the landing is through the ladder, the same rung spacing as used on the ladder shall be used from the landing platform to the first rung below the landing."

Although the reading material is dry, it is important for you to be aware of the regulations, because in the end, safety is your own concern. Your company, your direct supervisor, and your job foreman all are responsible for safety, but ultimately, safety is a personal responsibility. If you see a situation that you know does not meet 1910 regulations and standards, it is your responsibility to yourself and your fellow employees to report the violation. You can read more about this at the end of the chapter.

Other 1910 regulations deal with such issues as the following:

- **Confined spaces:** Confined spaces (as in tower legs and blades) are enclosed spaces that are large enough for workers to enter and perform work assignments but have limited openings through which to enter and exit and are not designed for continuous employee occupancy.

    Specific hazards of a confined space include one or more of the following characteristics:

    - Contains or has the potential to contain a hazardous atmosphere
    - Contains an engulfing potential
    - Contains a hazardous internal configuration
    - Contains other recognized safety or health hazards

    Workers and emergency responders can be injured because of the failure to recognize that a confined space is a potential hazard. Planned entry and rescue procedures are the two major components of a safe confined-space entry (League of Minnesota Cities, Risk Management Information, September 2008).

- **Lockout/tagout:** Workers performing service or maintenance on machinery and equipment are exposed to injuries from the unexpected energization—start-up—of the machinery or equipment or release of stored energy in the equipment. To protect such employees, the lockout/tagout standard requires the adoption and implementation of practices and procedures to shut down equipment, isolate it from its energy source(s), and prevent the release of potentially hazardous energy while maintenance and servicing activities are being performed. It contains minimum performance requirements and definitive criteria for establishing an effective program for the control of hazardous energy. However, employers have the flexibility to develop lockout/tagout programs that are suitable for their respective facilities. When a company uses lockout tags, it must provide additional training for all employees regarding the following limitations of tags (http://www.setonresourcecenter.com/safety/loto/):

○ Lockout tags are essentially warning devices and do not provide the physical restraint of a lock.

○ Lockout tags must be legible and understandable by all employees.

○ When a tag is attached to an isolating means, it is not to be removed except by the person who applied it, and it is never to be bypassed or ignored.

○ Tags and their attachments must be made of materials that will withstand the environmental conditions encountered in the workplace.

○ Tags may evoke a false sense of security. They are only one part of an overall energy-control program.

○ Tags must be securely attached to the energy-isolating devices so that they cannot be detached accidentally during use.

- **Fire regulations:** Here are a few parts of the 1910 regulations (http://www/osha.gov) that deal with fire prevention and response:

- **1926.150(a)(1):** "The employer shall be responsible for the development of a fire protection program to be followed throughout all phases of the construction and demolition work, and he shall provide for the firefighting equipment as specified in this subpart. As fire hazards occur, there shall be no delay in providing the necessary equipment."

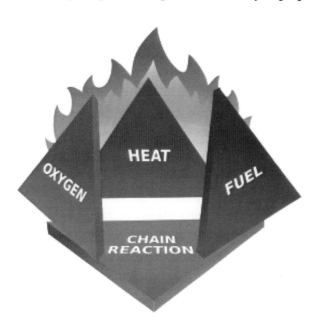

- **1926.150(a)(5):** "As warranted by the project, the employer shall provide a trained and equipped firefighting organization (Fire Brigade) to assure adequate protection to life."

- **1926.150(c)(1)(i):** "A fire extinguisher, rated not less than 2A, shall be provided for each 3,000 square feet of the protected building area, or major fraction thereof. Travel distance from any point of the protected area to the nearest fire extinguisher shall not exceed 100 feet."

These are just a few pieces of the OSHA 1910 regulations that you will learn in your company-sponsored OSHA 10-hour or 30-hour training sessions.

## *OSHA 10-Hour and 30-Hour Training*

In 1971, OSHA initiated an outreach-training program for general industry and construction industry safety training. The general industry courses follow the 1910 regulations (discussed earlier in this chapter), and construction follows the 1927 regulations. Employers are required to teach the regulations within prescribed outreach courses called the OSHA 10-Hour and the OSHA 30-Hour training. The 10-hour course, which takes approximately 1.5 days to complete, is a shorter version of the 30-hour course, which typically takes up 4.5-days. Because of time constraints and the job rescheduling required to accommodate training days, employers often send employees to the 10-hour training and managers to the more in-depth 30-hour course. Though these courses are required by OSHA, they are not taught by OSHA. As a rule, the courses are taught by a trainer who has completed the 40-hour trainer-authorization course provided by approved safety instructors, typically associated with state or professional safety associations.

The 10-hour training courses, both general industry and construction, may look something like this:

**Ten-Hour General Industry Training Topic Outline**

- 1 hour: introduction to OSHA, OSHA Act/General Duty Clause 5(a)(1), inspections, citations, and penalties (1903), record keeping (1904)
- 1 hour: walking and working surfaces (subpart D)
- 1 hour: means of egress and fire protection (subparts E and L)
- 1 hour: electrical (subpart S)
- 1 hour each of any three or more:
  - Flammable and combustible liquids (subpart H)
  - Personal protective equipment (subpart I)
  - Machine guarding (subpart O)
  - Hazard communication (subpart Z)
  - Introduction to industrial hygiene/bloodborne pathogens, and/or ergonomics (subpart Z)
  - Safety and health programs
- Not specified: maintenance personnel—ergonomics, hazard communication, and forklifts (if applicable)
- Not specified: office personnel—ergonomics and hazard communication
- Remainder of time: Any other general industry standards and policies and/or expansion of the required topics

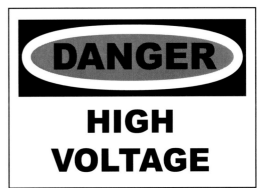

**Ten-Hour Construction Training Topic Outline**

- 1 hour: introduction to OSHA, OSHA Act/General Duty Clause 5(a)(1), general safety and health provisions, competent person (subpart C)
- 1 hour: electrical (subpart S)
- 1 hour: fall protection (subpart M)
- 1 or more topics at 1 hour each:
  - Materials handling, storage, use and disposal (subpart H)
  - Cranes, derricks, hoists, elevators, and conveyors (subpart N)
  - Motor vehicles, mechanized equipment, construction equipment (subpart O)
- 1 or more (as above):
  - Scaffolding (subpart L)
  - Excavations (subpart P)
  - Stairways and ladders (subpart X)
- Remainder of time: Any other construction industry standards and policies and/or expanded coverage of the required topics

**Thirty-Hour General Industry Training Topic Outline**
- 3 hours: introduction to OSHA, OSHA Act/General Duty Clause 5(a)(1), Inspections, Citations, and Penalties (1903)
- 2 hours: walking and working surfaces (subpart D)
- 2 hours: means of egress and fire protection (subparts E and L)
- 2 hours: flammable and combustible liquids (subpart H)
- 1 hour: personal protective equipment (subpart I)
- 2 hours: permit required confined spaces (subpart J)
- 2 hours: lockout/tagout (subpart J)
- 1 hour: materials handling (subpart N)
- 2 hours: machine guarding (subpart O)
- 1 hour: welding, cutting, and brazing (subpart Q)
- 2 hours: electrical and safety-related work practice (subpart S)
- 2 hours: hazard communication (subpart Z)
- 2 hours: introduction to industrial hygiene/blood-borne pathogens (subpart Z)
- 1 hour: record keeping (1904)
- 1 hour: ergonomics
- 1 hour: safety and health programs
- 3 hours: other applicable OSHA standards or policies or expanded coverage

of the required topics

**Thirty-Hour Construction Training Topic Outline:**

- 1 hour: introduction to OSHA, OSHA Act/General Duty Clause 5(a)(1)
- 1 hour: general safety and health provisions, competent person (subpart C)
- 1 hour: clarification of citation policy regarding 1926.20 and 1926.21 and related general safety and health provisions
- 1 hour: occupational health and environmental controls—emphasis on hazard communication (subpart D)
- 1 hour: health hazards in construction (subpart D)
- 1 hour: personal protective and lifesaving equipment (subpart E)
- 1 hour: fire protection and prevention (subpart F)
- 1 hour: materials handling, rigging, storage, use, and disposal (subpart H)
- 1 hour: hand and power tools; machine guarding (subpart I)
- 1 hour: welding and cutting (subpart J)
- 2 hours: electrical (subpart K)
- 1 hour: scaffolding (subpart L)
- 2 hours: fall protection (subpart M)
- 1 hour: cranes, derricks, hoists, elevators, and conveyors (subpart N)
- 1 hour: motor vehicles, mechanized equipment, and marine operations; rollover protective structures and overhead protection; and signs, signals, and barricades (subparts O, W, and G)
- 1 hour: excavations (subpart P)
- 1 hour: concrete and masonry construction (subpart Q)
- 1 hour: stairways and ladders (subpart X)
- 1 hour: confined-space entry
- 9 hours: expanded coverage of the required topics and/or supplementation of those topics coverage of other construction standards

Even though these courses and the 1910 and 1926 regulations are not specific to the wind energy industry, there are many applicable standards within these training courses that you will be able to translate to your everyday work situations. Remember, safety is job 1, and the hours you invest in learning the OSHA regulations are hours you are investing the protection of your own life, as well as that of your coworkers.

## *Additional Safety Considerations*

You will probably take a definitive safety course as part of your college studies, and the information presented in this chapter is meant to be only an introduction. In keeping with that goal, there are a few other areas of safety to explore here.

## Crane Safety

Cranes are necessary for much of the work that occurs on wind turbines. Whether setting new turbines in place or lifting workers to the nacelle, cranes play a vital yet dangerous role in wind turbine repair. Anytime you are working near a crane, you must constantly be aware of the crane's position. Although crane operators must be licensed and complete special OSHA training, you cannot depend on the crane operator or spotter to watch out for your safety. It cannot be emphasized enough that your safety is your own responsibility. You must stay out of the crane swing zone, which is located around the crane's superstructure. It is often marked off with a warning line running from the outriggers back to the frame of the crane. The area includes the operator's cab, the engine, and the counterweight. Anyone within this area is in danger of being crushed.

Cranes also are prohibited from operating within a ten-foot radius of any power lines, electrical poles, or other electrical hazards. All workers should be aware of the location of power lines and other electrical hazards in their work areas, so that if a crane is in danger of violating the safe zones, someone can alert the crane operator or spotter.

*Crane placing a Vestas V82 hub onto ground at TSTC campus*

A spotter is a person assigned to the crane operator to help watch that no safety zones are violated and to ensure that outriggers are placed correctly so that the crane will not tip over. Sometimes the crane company provides the spotter, but sometimes turbine repair technicians take on that responsibility. OSHA has very specific crane operation guidelines, and you may encounter sections of these in your 10-hour or 30-hour training courses.

**Extreme Weather**

Weather conditions can present a variety of hazards for those working with wind turbines. In northern climates, where winters are more severe, icing of the blades can be a problem. In, fact, it is such a serious problem that in a research article presented at the BOREAS (Boreal Ecosystem Atmosphere Study) conference in 2003 in Pyhä, Finland, and titled "Risk Analysis of Ice Throw from Wind Turbines," the authors reported that

> In Germany and Austria ice throw/fall prediction reports are required by the building authorities of some districts, especially in the inland and mountainous regions. Together with the increasing number of wind turbines at these sites, the number of ice-throw reports for building permission increases. It is to be expected that in connection with this, the number of experts and competing companies will increase as well and will improve the knowledge. As a general recommendation, it can be stated that wind farm developers should be very careful at ice endangered sites in the planning phase and take ice throw into account as a safety issue.

According to Patty Winsa, staff writer for TheStar.com:

> In cold climates, the right combination of moisture and cold air can cause ice buildup on blades. This not only reduces the aerodynamic performance of wind turbines, but ice thrown from a blade can travel "up to several hundred metres," according to GE Energy, which makes turbines. That's what happened in England. People near a wind farm in Lincolnshire reported mysterious glowing orbs streaking through the sky and thunder before one 20-metre blade was thrown to the ground and another was damaged. Experts blamed ice thrown from a turbine.

Another danger presented by extreme weather is lighting strike. On its Web site, LM Glasfiber, a maker of wind turbine blades that is headquartered in Denmark, states that "lightning strikes are a wind turbine's worst enemy. Without effective lightning protection, both the blades and the turbine itself can be severely damaged by the powerful energy surges in lightning." They go on to note that a direct lighting strike on a wind turbine blade can cause temperature spikes on the blade's surface of up to 30,000°C. The danger to humans working on wind turbines is obvious. AES Winds of Lake Benton, Minnesota, a subsidiary of AES Corporation (which is headquartered in Arlington, Virginia), provides electricity to 3,300 homes in southwestern Minnesota. Company spokespeople state that accurate weather forecasting is very important to their industry, and like most wind farms, AES provides lighting-arrest systems for all of its turbines and subscribes to high-tech weather forecasting systems. Said John Brown, safety coordinator technician, "When there is lightning, we notify crews in the field right away and they immediately come down from the towers" (http://www.dtn.com/files/case_studies/weather/ENERGY/CS_AESWINDS_1107.pdf).

Not only lightning but also hail, high winds, and tornadoes endanger work crews. Wind turbine technicians must remain alert to changing weather situations and respond quickly when advised of impending storms.

## *What If Your Boss Does Not Follow the Safety Rules?*

The safety precautions your employer puts in place are for your benefit as an employee. Although the majority of employers are diligent in ensuring the safety of their employees, you may on occasion encounter a supervisor or manager who is prone to cutting corners and taking shortcuts. If at any time you think that your employer is not following safety standards, it is your responsibility to yourself and your coworkers to speak out. Here are a few of the rights that you, as an employee, have under OSHA guidelines (*Taking Charge: Your Education, Your Career, Your Life*, Karen Smith, 2008):

- To review employer-provided OSHA standards, regulations, and requirements
- To request information from the employer on emergency procedures
- To receive adequate safety and health training when required by OSHA standards regarding toxic substances and any procedures set forth in emergency action plans required by an OSHA standard
- To ask the OSHA area director to investigate hazardous conditions or violations of standards in your workplace
- To file anonymous complaints with OSHA
- To be advised of OSHA actions regarding your complaint and to have an informal review of any decision not to inspect or to issue a citation
- To have your employee representative accompany the OSHA compliance officer on inspection
- To seek safe and healthful working conditions without your employer retaliating against you
- To observe any monitoring or measuring of toxic substances or harmful physical agents and review any related monitoring or medical records

If you believe your employer or supervisor is not taking proper precautions, you should alert someone in authority above your position. If your immediate supervisor appears to be at fault, you will have to be prepared to go over his or her head. If you do not get satisfaction, keep going up the ladder of authority until someone listens. OSHA fines companies for regulatory noncompliance, so your employer will most likely be willing to listen to any safety concerns. Being in the position of whistleblower is never comfortable, but when you are working 260 feet off the ground, lives are always at stake.

# *Conclusion*

You have mostly like heard the saying "Safety is job 1." Safety really must be your first concern as a wind turbine technician: Your life and the lives of those around you depend on every action you take on the job. To help you prepare to work safely, this chapter taught you about why the Occupational Safety and Health Administration (OSHA) exists, why safety is your responsibility, what personal protective equipment is and why you should use it, what the two-person rule is and why it is important, why you need to learn first-aid techniques and cardiopulmonary resuscitation, the topics covered in OSHA safety courses, crane safety, extreme-weather safety, and why and how to be a whistle-blower.

# *References and Additional Resources*

### Lockout/Tagout
http://www.setonresourcecenter.com/safety/loto/

### Suspension Trauma
http://en.wikipedia.org/wiki/Suspension_trauma

### Confined-Space Entry
League of Minnesota Cities. Risk management information. September 2008. Available from http://www.lmnc.org/media/document/1/confinespaceentry.pdf. Accessed March 14, 2008.

### Extreme-Weather Dangers
Seifert, Henry; Westerhellweg, Annette; Kröning, Jürgen; DEWI (German Wind Energy Institute). Risk analysis of ice throw from wind turbines. Paper presented at BOREAS VI, April 9–11, 2003, Pyhä, Finland. Available from http://web1.msue.msu.edu/cdnr/icethrowseifertb.pdf Accessed July 8, 2009.

Winsa, Patty. Falling ice at perial at wind farms. TheStar.com, February 2, 2009. Available from http://www.thestar.com/News/GTA/article/580767. Accessed March 15, 2009.

Effective lightning protection. LM Glasfiber Web site. http://www.lmglasfiber.com/Products/Lightning.aspx. Accessed July 8, 2009.

### OSHA
http://www.osha.gov

# *Review Questions*

1. What does "Safety is job 1" mean? How will you apply this concept on the job?

2. Explain the concept of total risk management as it applies to working on wind turbines.

3. What is PPE? What unique pieces will you specifically need on the job?

4. What is the two-person rule? Why is it important to always work as part of a team?

5. Describe the effects of suspension trauma.

6. What is OSHA? How does OSHA pertain to your career as a wind energy technician?

7. What is the purpose of OSHA's 1910 regulations?

8. List five things you will learn in the OSHA 10-hour training

9. What do the terms swing zone and swing radius refer to when dealing with crane safety?

10. Why is extreme weather a concern for those working on wind turbines? What type of weather must you watch for?

## *Application Exercises*

1. Using the Internet, research the OSHA 1910 regulations. From the information you find, describe the process of lockout/tagout. When might you use this during wind turbine repair?

2. Research the effects of extreme weather on wind turbines, using the links at the end of this chapter as well as at least two other outside sources. From your research, answer the following questions:

   a. What type of weather are you most likely to encounter if you work in Texas?

   b. Describe the effects of lightning on wind turbine blades and mechanisms.

   c. Why is ice buildup on the blades a problem? In what areas of the United States are you most likely to encounter this phenomenon?

3. Role-playing: Working in groups of three to four people, devise a scenario in which your boss expects you to cut corners during wind turbine repair. Demonstrate during a skit how you would (a) approach your boss and (b) approach his or her supervisor.

## Chapter Three
# Fasteners

## Learning Objectives

*As a wind turbine repair technician, you will quickly discover that because various companies from around the world make turbines, each company has its own methods and conventions for building equipment. You can best deal with those differences by having an in-depth understanding of the various types of fasteners you will encounter.*

*In this chapter, you will learn about:*

- *The different types of fasteners, including bolts, screws, nuts, and washers*
- *The types of threads and various classes of fits of fasteners*
- *The types of materials used to create fasteners*
- *The various stresses in strength applied to fasteners and to anticipate how these outside factors affect fasteners*

## *Introduction*

All threaded fasteners are made of a helical structure used to convert between rotational and linear movements or forces. When a fastener is tightened, it is similar to driving a wedge into a gap so that it sticks tightly. Usually the pitch of a screw thread is such that friction is sufficient to "prevent linear motion being converted to rotary movement" (http://en.wikipedia.org/wiki/Screw_thread).

Matched pairs of threads, both internal and external, are described as having male and female gender. A screw, for example, is the male part, whereas the matching receiving end has female threads. Fasteners may also be left- or right-handed. The majority of fasteners are oriented so that they are tightened by moving them in a clockwise, or right-handed, direction. This is called the right-hand rule and is easily remembered with the saying "Righty-tighty, lefty-loosey." Threads that tighten counterclockwise are considered to have a left-hand grip.

Although most fasteners are right-hand threaded by default, left-hand threads may be used in the following situations:

- Where the rotation of a shaft would cause a conventional right-handed nut to loosen rather than to tighten because of fretting-induced precession, as on a left-hand bicycle pedal
- In combination with right-handed threads in turnbuckles

- In some gas-supply connections to prevent dangerous misconnections—gas welding, for example—the flammable gas supply uses left-handed threads
- In some instances—for example, early ballpoint pens—to provide a "secret" method of disassembly
- In some applications of a leadscrew—for example, the cross slide of a lathe, where it is desirable for the cross slide to move away from the operator when the leadscrew is turned clockwise

## *Diameter and Pitch*

Screw threads come in several relevant diameters. The *major diameter* refers to the largest diameter of a screw thread, whereas the *minor diameter* refers to the smallest diameter of a screw thread. Pitch is the distance between an adjacent thread measured parallel to the thread axis.

# Basic Profile of the Unified and ISO Thread Form

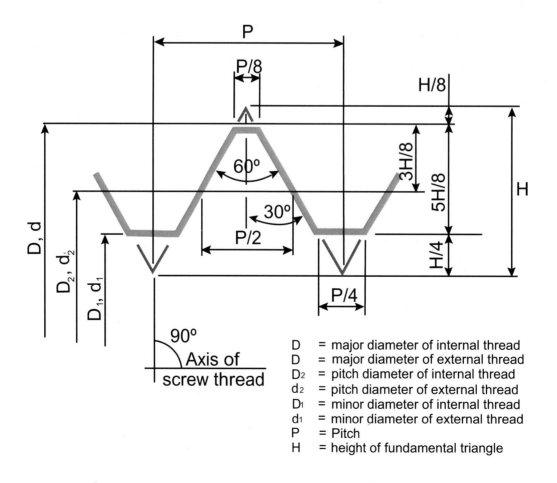

D  = major diameter of internal thread
D  = major diameter of external thread
$D_2$ = pitch diameter of internal thread
$d_2$ = pitch diameter of external thread
$D_1$ = minor diameter of internal thread
$d_1$ = minor diameter of external thread
P  = Pitch
H  = height of fundamental triangle

## *Types of Fasteners*

As a wind turbine repair specialist, you are likely to encounter all of the different types of fasteners. In this section you will find a partial list of fasteners, along with helpful diagrams. You will see that there are both bolts and screws in the list. It is important to remember that a screw and a bolt are not the same type of fastener. Typically, a screw is threaded all the way to the head, whereas a bolt has a short shank between the head and the threads. However, this is not always the case. One differentiation is that a bolt is designed to be used with a corresponding nut.

- **Hex bolts** are externally threaded fasteners designed for insertion through holes in assembled parts and are normally intended to be tightened or released by torquing a nut. ANSI B18.2.1 (http://webstore.ansi.org/ RecordDetailaspx?sku=ANSI%2FASME+B18. 2.1-1996+(R2005), the code of standards from the American National Standards Institute regarding square and hex bolts and screws sized by inches, lists hex bolts from ¼" through 4" diameter and heavy hex bolts from ½" through 3".

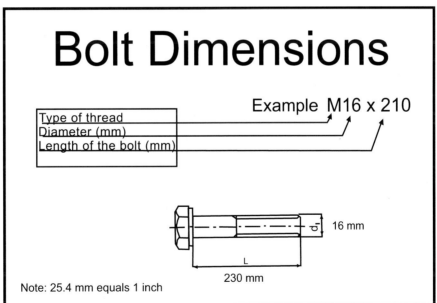

# Bolt Dimensions

Example M16 x 210

Type of thread
Diameter (mm)
Length of the bolt (mm)

$d_1$   16 mm

L

230 mm

Note: 25.4 mm equals 1 inch

- Cap screws are externally threaded fasteners capable of insertion through holes in assembled parts. They mate with a preformed internal thread in one of those parts; normally they are tightened and released by torquing the head.

- **Machine screws** are externally threaded fasteners capable of insertion through holes in assembled parts. They mate with preformed threads in one of the parts or with a nut. They are most popular in sizes from $\frac{2}{56}''$ through $\frac{1}{4}''$. There are a variety of head styles available for diverse applications.

- **Self-tapping screws** are externally threaded fasteners designed to be capable of insertion through holes in assembled parts, forming a matching internal thread in one of the parts. They are tightened and released by torquing the head.

- **Nuts** are the internally threaded fastener designed to assemble with bolts or other externally threaded parts.

- **Flat washers** are used to increase the bearing surface and to reduce friction.

- **Lock washers** provide a measure of resistance to "backing off."

- **Studs (threaded rod)** are similar to bolts but-without heads. Studs are threaded on both ends. In some cases the entire length of the stud is threaded; in other cases there will be a nonthreaded section in the middle.

Threaded fasteners come in both coarse and fine threads. The coarse-thread series UNC/UNRC is the most commonly used system. It is used for producing threads in low-strength materials such as cast iron, mild steel, softer copper alloys, and aluminum. The coarse thread is also used for rapid assembly or disassembly.

Fine threads, the UNF/UNRF series, are used for applications that require a higher tensile strength than the coarse-thread series and where a thin wall is required. The extra-fine-thread series, UNEF/UNREF, is used when the length of engagement is smaller than the fine-thread series. It is also applicable in all situations calling for a fine thread.

## Classes of Threads

Thread classes are distinguished from each other by the amounts of tolerance and allowance specified. For example, class would come into play when designing a component. The looser the fit, the weaker the design, but the part would also be cheaper to make. External threads or bolts are designated with the suffix *A*, whereas internal, or nut, threads are designated by *B*. For more information, refer to http://www.sizes.com/tools/bolts_inch_threadFit.htm. The following is a list of class threads:

- **Grades 1A and 1B:** For works of rough commercial quality where loose fit for spin-on-assembly is desirable
- **Grades 2A and 2B:** The recognized standard for normal production of the great bulk of commercial bolts, nuts, and screws
- **Grades 3A and 3B:** Used for a close fit between mating parts when high-quality work is required
- **Grade 4:** A theoretical rather than practical class, now obsolete
- **Grade 5:** For a wrench set or what is called interference fit; used principally for studs and their mating tapped holes. A forced fit requiring the application of high torque for semipermanent assembly, fasteners with this grade of threads typically are not expected to ever be removed.

K-T Bolt Manufacturing, Inc. offers the following advice on material choice on its Web site (http://www.k-tbolt.com/fastener_materials.html):

> In addition to being graded, fasteners are manufactured in a wide range of materials, from common steel to titanium, plastic, and other exotic materials. Many materials are further separated into different grades to describe specific alloy mixtures, hardening processes, etc. Moreover, some materials are available with a variety of coatings or plating to enhance the corrosion resistance, lubrication, or appearance of the fastener.

Because all of these differences can affect such factors as strength, brittleness, and corrosion, K-T Bolt advises that it is always wisest to replace a bolt with the exact same material when making repairs.

Steel is the most common fastener material. Steel fasteners are available plain as well as with various surface treatments, such as zinc plating, galvanization, and chrome plating.

Typically, bolts of different grades have marking on their heads to identify the grade of fastener (charts). This will help the grades of fasteners that you need for steel.

Steel fasteners are commonly available in four grades. Many other grades exist but are used far less often. The most common grades are grade 2, grade 5, grade 8, and alloy steel. Fasteners that are grade 2, 5, or 8 are usually plated with a silver or yellow zinc coating or are galvanized to resist corrosion.

- **Grade 2:** A standard hardware grade steel. This is the most common grade of steel fastener and is the least expensive. Grade 2 bolts have no head marking, although sometimes a manufacturer's mark is present.
- **Grade 5:** Hardened to increase strength, grade 5 fasteners are the most common bolts found in automotive applications. Grade 5 bolts have three evenly spaced radial lines on the head.

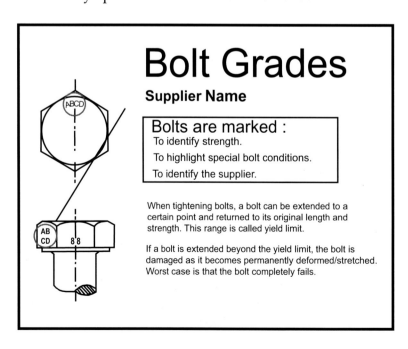

## Metric Mechanical-Property Classes for Steel Bolts, Screws, and Studs

| Head Marking | Property Class | Size Range, Inclusive | Material | Minimum Proof Strength (MPa) | Minimum Tensile Strength (MPa) | Minimum Yield Strength (MPa) |
|---|---|---|---|---|---|---|
| 4.6 | 4.6 | M5–M36 | Low- or medium-carbon steel | 225 | 400 | 240 |
| 4.8 | 4.8 | M1.6–M16 | Low- or medium-carbon steel | 310 | 420 | 349 |
| 5.8 | 5.8 | M5–M24 | Low- or medium-carbon steel | 380 | 520 | 420 |
| 8.8 | 8.8 | M16–M36 | Medium-carbon steel, Q&T | 600 | 830 | 660 |
| 9.8 | 9.8 | M1.6–M16 | Medium-carbon steel, Q&T | 650 | 900 | 720 |
| 10.9 | 10.9 | M5–M36 | Low-carbon martensite steel, Q&T | 830 | 1040 | 940 |
| 12.9 | 12.9 | M1.6–M36 | Alloy steel, Q&T | 970 | 1220 | 1100 |

Q&T = quenched and tempered.

- **Grade 8:** Have been hardened more than grade 5 bolts and thus are stronger and are used in demanding applications such as automotive suspensions. Grade 8 bolts have six evenly spaced radial lines on the head.
- **Alloy steel:** These fasteners are made from a high-strength steel alloy and are further heat treated. Alloy steel bolts are typically not plated, resulting in a dull black finish. Alloy steel bolts are extremely strong but very brittle.
- **Stainless steel:** An alloy of low-carbon steel and chromium for enhanced corrosion characteristics. Stainless steel is highly corrosion resistant for the price, and because the anticorrosive properties are inherent to the metal, it will not lose this resistance if scratched during installation or use. It is a common misconception that stainless steel is stronger than regular steel. In

fact, because of its low carbon content, stainless steel cannot be hardened. Therefore, when compared with regular steel, it is slightly stronger than an unhardened (grade 2) steel fastener but significantly weaker than hardened steel fasteners. Stainless steel is also much less magnetic than regular steel fasteners, though some grades will be slightly magnetic.

- **Aluminum:** The least costly, by volume, of all fastener metals. Aluminum fasteners are classified as hardened or nonhardened and weigh about one third as much as steel. Some grades equal or even exceed the tensile strength of mild steel. Aluminum is nonmagnetic, can be hardened by alloying, and has high corrosion resistance.

- **Bronze:** A lower-strength material that has excellent corrosion-resistance properties for seawater applications.

- **Brass:** Worked easily into shape and has adequate strength. The tensile strength, or hardness, of brass is improved by cold working. Some brasses have a greater tensile strength than mild carbon steel, along with a higher resistance to corrosion.

- **Monel:** Trade name for a nickel–copper alloy with high strength and excellent corrosion resistance in a range of media, including seawater, hydrofluoric acid, sulfuric acid, and alkalis. Used for marine engineering, chemical and hydrocarbon processing equipment, valves, pumps, shafts, fittings, fasteners, and heat exchangers.

- **Titanium:** Fasteners made from this material are used chiefly on aircraft. Titanium has excellent corrosion resistance and good high-temperature performance. These fasteners are most commonly used in joints loaded in shear but are also used in tension-loaded joints.

- **Inconel:** Trade name for a group of nickel-chromium-based superalloys that are excellent for fasteners that must retain high strength and oxidation resistance at temperatures up to 1600°F.

## *Bolt Strength*

To be able to choose the right fastener for a particular situation, you must know what kind of stress the materials to be fastened are under and what kind of strength each fastener material has.

- **Bolt strength:** Refers to the material's ability to resist an applied stress. The applied stress may be (a) tensile, (b) compressive, or (c) shear. Uniaxial stress is expressed as follows:

$$\sigma = \frac{F}{A}$$

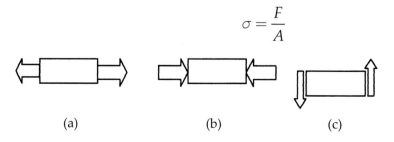

(a)                    (b)                    (c)

where *F* is the force (N, lbf) acting on an area, *A* (m², ft²). The area can be the undeformed area or the deformed area, depending on whether engineering stress or true stress is used.

- **Tensile stress:** Loading that tends to produce stretching of a material by the application of axially directed pulling forces. Any material that falls into the "elastic" category can generally tolerate mild tensile stresses, but materials such as ceramics and brittle alloys are very susceptible to failure under the same conditions. If a material is stressed beyond its limits, it will fail. The failure mode, either ductile or brittle, is based mostly on the microstructure of the material. Some steel alloys are examples of materials with high tensile strength.

- **Compressive stress, or compression:** The stress state when the material (compression member) tends to compact. A simple case of compression is the uniaxial compression induced by the action of opposite, pushing forces. Compressive strength for materials is generally higher than tensile stress, but geometry is very important in the analysis, because compressive stress can lead to buckling.

- **Shear stress:** Caused when a force is applied to produce a sliding failure of a material along a plane that is parallel to the direction of the applied force. An example is cutting paper with scissors.

- **Yield strength:** The lowest stress that gives permanent deformation in a material. In some materials, such as aluminum alloys, the point of yielding is hard to define, so it is usually given as the stress required to cause 0.2% plastic strain.

- **Compressive strength:** A limit state of compressive stress that leads to compressive failure in the manner of ductile failure (infinite theoretical yield) or in the manner of brittle failure (rupture as the result of crack propagation, or sliding along a weak plane—see *shear strength*).

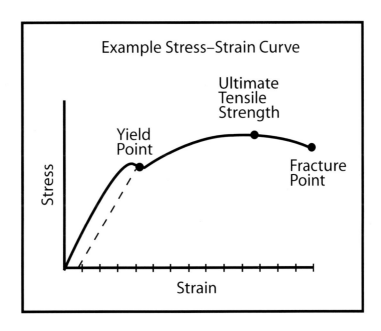

- **Tensile strength, or ultimate tensile strength:** A limit state of tensile stress that leads to tensile failure in the manner of ductile failure (yielding as the first stage of failure, some hardening in the second stage, and breaking after a possible "neck" formation) or in the manner of brittle failure (sudden breaking into two or more pieces with a low-stress state). Tensile strength can be given as either true stress or engineering stress.

- **Fatigue strength:** A measure of the strength of a material or a component under cyclic loading and is usually more difficult to assess than static strength. Fatigue strength is given as stress amplitude or stress range ($\Delta\sigma = \sigma_{max} - \sigma_{min}$), usually at zero mean stress, along with the number of cycles to failure.

# *Torque*

*Torque* is a measurement of how much a force acting on an object causes that object to rotate. The object rotates about an axis, which we will call the pivot point and will label O. We will call the force F. The distance from the pivot point to the point where the force acts is called the moment arm and is denoted by r. Note that this distance, r, is also a vector and points from the axis of rotation to the point where the force acts.

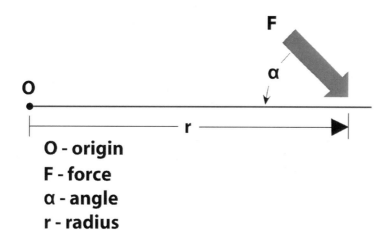

**O - origin**
**F - force**
**α - angle**
**r - radius**

# *Sample Problem*

If the amount of force that is applied to an object is 12 lbf and the radius of the torque wrench arm is 3.25 ft, determine the torque in feet–pound force (ft-lbf) and newton meters (N·m).

Given:

1 lbf = 4.448 N
1 ft = 0.3048 m
F = 12 lbf
r = 3.25 ft

A torque wrench sets the torque of a fastener and is used where tightness is crucial and must be exact. The wrench allows the operator to measure the amount of rotational force applied to the bolt in order to meet specifications. There are several types of torque wrenches available, and the choice of which type to use is based on application needed.

- **Deflecting beam type:** The beam torque wrench is the simplest to use and consists of a long lever arm between the handle and head. The arm material is flexible and bends somewhat under applied force. An indicator is attached to a second, smaller bar, which is connected to the head, parallel to the lever arm. The second arm does not flex and carries no strain. A calibrated scale is fitted to the handle, and bending the handle causes the scale to move, indicating torque. When the specified torque is reached, the operator releases the pressure. This is the most common type of torque wrench and is considered to be a "normal use" torque wrench, for applying pressure of up to 650 lb/ft.

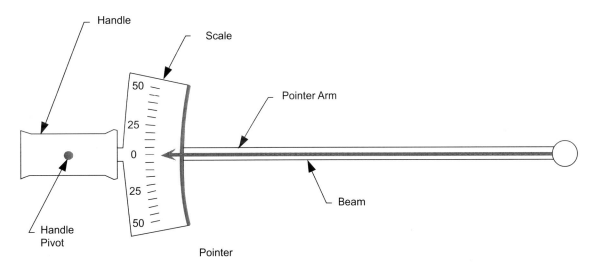

- **Dial Indicating:** This wrench applies torque to a coil rather than to a deflecting beam. Not only does this principle give the wrench a longer life span but it also employs a greater safety margin on maximum loading. Additionally, this wrench provides more consistent and accurate readings throughout the range of each wrench.

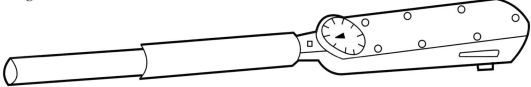

## DIAL INDICATING

- **Click type:** A calibrated clutch system, such as is the click torque wrench, provides a more sophisticated method of presetting torque. When the preselected torque is reached, the clutch slips, thus preventing overtightening.

Commonly, this wrench system uses a ball detent and spring, with the spring preloaded by an adjustable screw thread, calibrated in torque units. The ball detent transmits force until the preset torque is reached, at which point the force exerted by the spring is overcome and the ball "clicks" out of its socket. However, a number of variations of this design exist for different applications and different torque ranges.

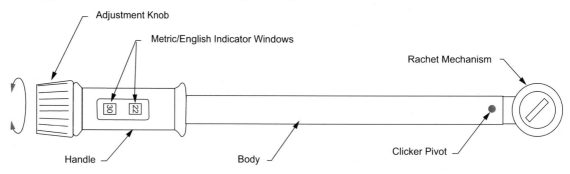

- **Electronic and programmable electronic type:** These are the most sophisticated of torque wrenches. Measurement is achieved by a strain gauge attached to the torsion rod. A transducer converts the signal generated to a the preselected unit of force, which when achieved shows on the digital display. The stored data can be transferred to a computer via the interface or sent to a printer.

- **High torque:** High-torque wrenches are designed to apply 650 ft-lbf of pressure and up. These are typically hydraulic wrenches or torque multipliers, which use a planetary gear design to achieve higher torque. Additionally, a high-torque wrench is used to check the bolts on the outside of the tower, which become loose over time as the wind exerts pressure against the tower.

The high-torque wrench uses hydraulics and pneumatics, making it inherently more technical in nature. The other torque wrenches described here are used for daily maintenance; the high-torque wrench is used in unique situations.

## *Conclusion*

All wind turbines are not alike, because there are several companies worldwide who make them, each with its own manufacturing specifications. You may be called on to repair quite a few models of turbines, so it is imperative that you be familiar with the various types of fasteners used in them. In this chapter, you learned about types of bolts, screws, nuts, and washers; types of fastener threads and classes of fit of fasteners; the kinds of materials that fasteners are made from; what types of stresses are put on fasteners; and how to calculate stress.

## *References and Additional Resources*

### Information on Fasteners

http://en.wikipedia.org/wiki/Screw_thread

http://www.portlandbolt.com/technicalinformation/thread-pitch.html

http://www.boltscience.com/pages/glossary.htm

### Information on Material Choice

http://www.k-tbolt.com/fastener_materials.html

http://www.thomasnet.com/articles/hardware/fastener-materials

## *Review Questions*

1. What is the difference between a screw and a bolt?

2. What is the pitch of a screw?

3. Name three situations in which left-handed threads are used.

4. How are thread classes distinguished from one another?

5. What is tolerance in regard to fasteners?

6. Why is material choice in fasteners important?

7. Describe the difference between tensile stress and compressive stress.

8. What is torque?

9. Describe two types of torque wrenches you will use on the job.

10. Describe three types of materials used to make fasteners and tell where they are most likely used.

## *Application Exercises*

1.  Determine the amount of force applied (in Newtons) if the final torque is 2500 ft-lbf and the radius of the torque wrench arm is 5 ft. Does this value seem reasonable for a normal person to achieve?

    Given:

    1 lbf = 4.448 N
    1 ft = 0.3048 m
    $\tau$ = 4.448 ft-lbf
    r = 5 ft

2.  Working up tower inside the nacelle, you are reassembling the coupling between the gearbox and generator. The flange is secured using twelve 1-8UNC SAE grade 5 bolts. As you install the fasteners, you can locate only eleven fasteners in the work area. A coworker states that there are spare fasteners down tower that you can use. When you get down tower, you discover a box of fasteners not properly marked. The following bolts are available in the box:

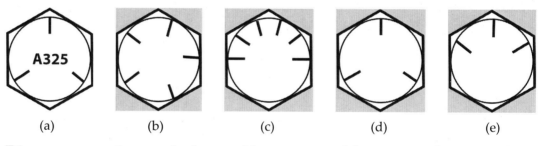

    (a)            (b)            (c)            (d)            (e)

    Discuss your actions and what problems you could encounter because of these actions.

## Chapter Four

# Gearboxes, Gear Ratios, and Failure Mechanisms

## Learning Objectives

*Energy created by wind impinges, or acts, on turbine blades, causing them to rotate. Therefore, it is imperative to have a thorough understanding of the mechanisms that a wind turbine uses to convert wind energy into electric energy.*

*In this chapter, you will learn:*

- *What a gear is and how to recognize the types of gears*
- *Specific gears for wind turbines*
- *The components of a gearbox*
- *Speed ratios*
- *What causes gears to fail*
- *How to lubricate and inspect gears*

## *Introduction*

Gears, most prominently located within the gearbox, are widely used in turbines. Rotation is transferred through the hub to the main shaft, which consists of four primary components: the shaft, the bearing, the trunnion mounts, and the gearbox. They provide mechanical advantages, such as transferring torque from one shaft to another. For example, the blades of a turbine rotate at a very slow speed and produce high torque, but the generator, to produce power, needs high speed at a low torque. The gearbox transmits the relatively low rotational speed of the hub to a higher speed necessary for the generator to create electricity.

## *The Basics*

### What Is a Gear?

A gear is a wheel with teeth around its circumference. Those teeth mesh with similar teeth on another mechanical device. Sometimes that is another gear wheel, so that force can be transmitted between two devices. A gear can mesh with anything that has teeth compatible with the gear's teeth, such as a rack or a nonrotating part. The most common situation is for one gear to mesh with another gear. Although gears are sometimes

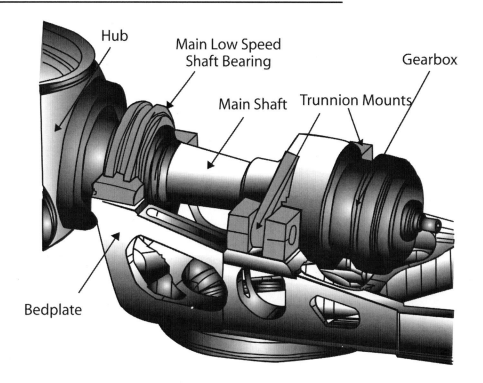

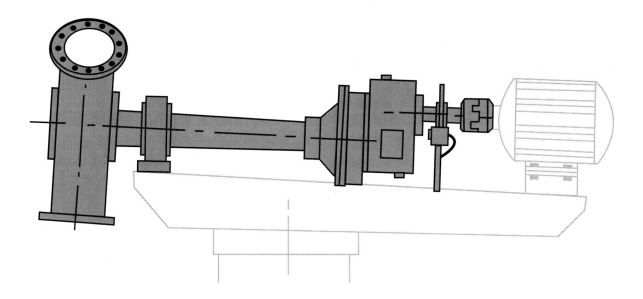

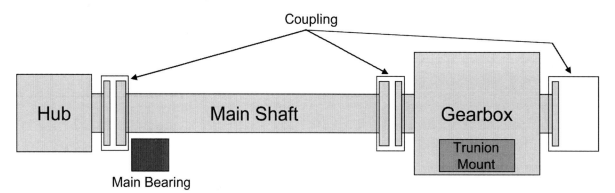

*Standard components of gear train inside nacelle*

used simply to transmit rotational forces from one shaft to another, perhaps their most important feature is that when the gears are of unequal sizes (diameters), a mechanical advantage is achieved. This means that the rotational force (torque) and rotational speed of the second gear is different from that of the first. This size difference allows gears to provide a means of increasing or decreasing rotational speed or torque. Such is the case in a wind turbine, where a smaller gear meshes with the larger gear to lower torque and create the higher speed necessary to create energy.

### Gear Terminology

- **Pitch circle:** The pitch circle is the circumference of a hypothetical smooth gear, or a gear that has barely visible teeth. The diameter of the pitch circle is known as the pitch diameter (d). With teeth of finite size, some of each tooth will extend beyond the pitch circle and some of it below the pitch circle.

- **Tooth face:** The face of the tooth is the location where the face of a corresponding tooth on another gear will meet.

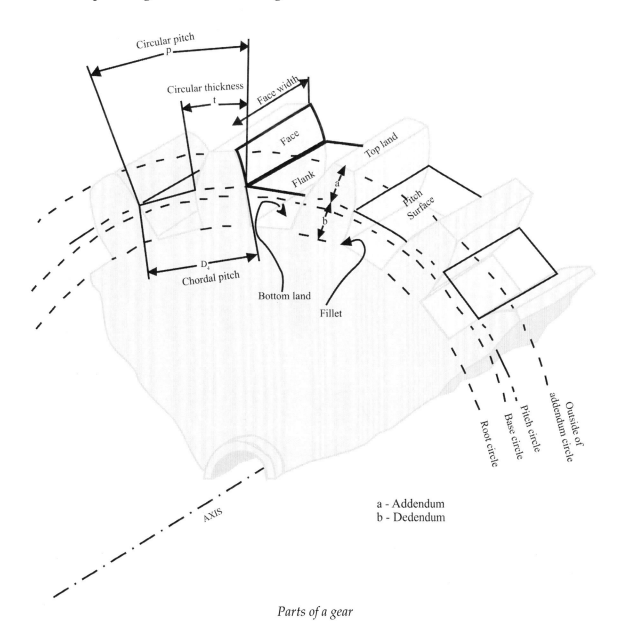

*Parts of a gear*

- **Face width:** The width of the face is the dimension parallel to the gear's axis of rotation.

- **Circular pitch:** The circular pitch (p) of a gear is the distance from one face on one tooth to the face on the same gear of the next tooth around the pitch circle.

- **Backlash:** Backlash is the free space between the gear teeth when they mesh together. When gears are manufactured, they are cut a little smaller than they need to be, resulting in the free space between the gear teeth. Backlash can be increased by moving the gears farther apart. Backlash can also be thought of as the error in motion when you change a gear's direction. For example, if you put your car into park without setting the emergency brake, the resulting roll is backlash, which causes wear on your car's gears.

### Types of Gears

- **Spur gears:** Spur gears are the most commonly used type. Tooth contact is primarily rolling, with sliding occurring during engagement and disengagement. Some noise is normal, but it may become objectionable at high speeds.

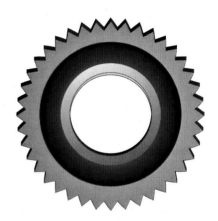

- **Rack and pinion:** Rack and pinion gears are essentially linear-shaped versions of spur gears. The spur rack is a portion of the spur gear with an infinite radius.

- **Internal ring gear:** An internal ring gear is a cylindrical gear with the meshing teeth inside or outside a circular ring. Often used with a spur gear, internal ring gears may be used within a planetary gear arrangement.

- **Helical gear:** A helical gear is a cylindrical gear that operates with less noise and vibration than spur gears. At any time, the load on helical gears is distributed over several teeth, resulting in reduced wear. Owing to their angular cut, teeth meshing results in thrust loads along the gear shaft. This action requires thrust bearings to absorb the thrust load and maintain gear alignment. They are widely used in industry. A negative is the axial thrust force that the helix form causes.

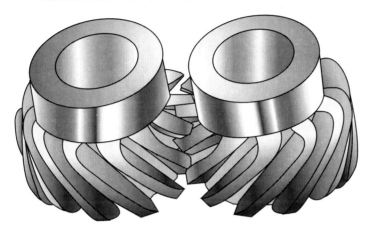

- **Helical rack gears:** Helical gears are linear-shaped gears that mesh with a rotating helical gear.

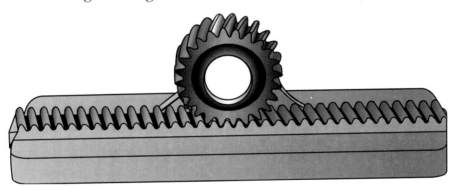

- Double helical gear: A double helical gear may have both left-hand and right-hand helical teeth. The double helical form is used to balance the thrust forces and provide additional gear shear area.

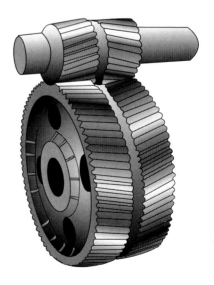

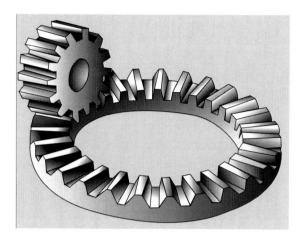

- **Face gears:** Face gears are circular discs with a ring of teeth cut on one side. The gear teeth are tapered toward the center of the tooth. These gears typically mate with a spur gear.

- **Worm gears:** Worm gears teeth resemble Acme screw threads, which mate with a helical gear, except that the helical gear is made to envelope the worm as seen along the worm's axis. The operation of worm gears is analogous to that of a screw. The relative motion between these gears is sliding rather than rolling. The uniform distribution of tooth pressures on these gears enables the use of metals with two inherent coefficients of friction, such as bronze wheel gears with hardened-steel worm gears. These gears rely on full fluid film lubrication and require heavy oil compounded to enhance lubricity and film strength to prevent metal contact.

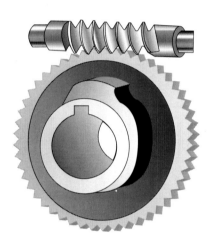

- **Hypoid gears:** Hypoid gears are typically found within the differential (rear axle) of automobiles. The gear arrangement allows the translation of torque 90 degrees. Hypoid gears are similar to spiral bevel gears except that the shaft centerlines do not intersect. Hypoid gears combine the rolling action and high tooth pressure of spiral bevels with the sliding action of worm gears. This combination and the all-steel construction of the drive and driven gear result in a gear set with special lubrication requirements, including oiliness and antiweld additives to withstand the high tooth pressures and high rubbing speeds.

- **Straight bevel gears:** Straight bevel gears have tapered conical teeth that intersect the same tooth geometry. These gears are used to transmit motion between shafts with intersecting center-lines. The intersecting angle is normally 90 degrees but may be as high as 180 degrees. When the mating gears are equal in size and the shafts are positioned at 90 degrees to each other, they are referred to as miter gears. The teeth of bevel gears can also be cut in a curved manner to produce spiral bevel gears, which produce smoother and quieter operation than straight-cut bevels do.

- **Spiral bevel gears:** Spiral bevel gears have helical-angle spiral teeth.

- **Screw gears:** Screw gears are helical gears of opposite helix angles that will mesh when their axes are crossed.

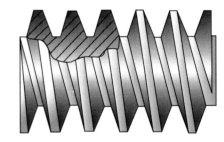

## *Types of Gearboxes*

A gearbox is a piece of equipment that houses gears in an arrangement suitable for changing speed or torque (or both). The primary gearbox you will encounter in the wind energy industry is the planetary arrangement. The main components of this type of gearbox are

- An interior-toothed gear wheel (ring gear)
- Two or three smaller toothed gear wheels (planet gear)
- A common carrier arm (planet carrier)
- A centrally placed toothed gear wheel (the sun gear)

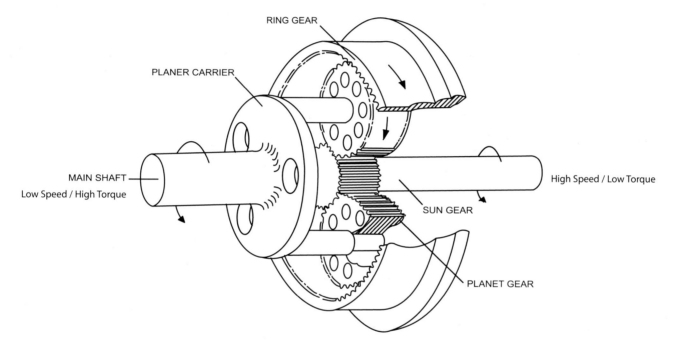

The advantages of using a planetary gearbox makes this the arrangement of choice for most wind turbine manufacturers. The advantages are as follows:

- Increased efficiency and extremely low speeds
- High reduction ratios and a higher torque
- Compactness and light weight; reduced requirement for installation space
- High reliability because of proper distribution of stress among different bearing components

Gearboxes are used in other places throughout industry in general, including power, chemical processing, and submarines. Here are a few of the types you may encounter:

- Bevel gearbox
- Helical gearbox
- Sequential gearbox
- Crane-duty gearbox

- Spiral-bevel gearbox
- Worm-reduction gearbox
- Cycloidal gearbox
- Offset gearbox
- Right-angle-bevel gearbox
- Shaft-mounted gearbox
- Worm gearbox

## *Gear Functionality*

The speed at which gears rotate is measured in revolutions per minute, abbreviated as **rpm**, **RPM**, **r/min**, or $\mathbf{r \cdot min^{-1}}$. The rpm is a unit of frequency—that is, the number of full rotations completed in 1 minute around a fixed axis. It is most commonly used as a measure of rotational speed or angular velocity of some mechanical component.

The corresponding Système International (SI) unit would be the hertz, which converts this way:

$$3600 \text{ r/min} \bullet 60 \text{ revolutions per second} \bullet 60 \text{ Hz}$$

With the SI, you will often use the unit for angular velocity, which is *radians per second* ($\text{rad} \bullet s^{-1}$):

$$1 \text{ r/min} \bullet 2\pi \text{ rad} \bullet min^{-1} \bullet \frac{2\pi}{60} \text{ rad} \bullet s^{-1} \bullet 0.10471976 \text{ rad} \bullet s^{-1}$$

To convert revolutions per minute to revolutions per second (hertz), you simply divide by 60. The opposite is true when converting from hertz to rpm, where you multiply by 60 instead.

Gear ratio is the relationship between the number of teeth on one gear and the number of teeth on another gear that mesh together, between the numbers of teeth on each of two sprockets that are connected with a common roller chain, or between the numbers of teeth on the circumferences of each of two pulleys connected with a drive belt. In the illustration, the smaller gear (known as the pinion) has 13 teeth, whereas the second, larger gear (known as the idler gear) has 21 teeth. The gear ratio is therefore 13/21 or 1/1.62 (also written as 1:1.62).

This means that for every one revolution of the pinion, the gear has made 1/1.62, or 0.62, revolutions. In practical terms, the gear turns more slowly than the pinion.

Suppose the largest gear in the picture has 42 teeth. Then the gear ratio between the second and third gear is 21/42 = 1/2. For every revolution of the smallest gear, the largest gear has only turned 0.62/2 = 0.31 revolution, a total reduction of about 1:3.23.

The intermediate (idler) gear directly contacts both the smaller and the larger gear, so it can be removed from the calculation, also giving a ratio of 42/13 = 3.23.

Because the number of teeth is also proportional to the circumference of the gear wheel (the bigger the wheel, the more teeth it has), the gear ratio can also be expressed as the relationship between the circumferences of both wheels (where $d$ is the diameter of the smaller wheel and $D$ is the diameter of the larger wheel):

$$\text{gr} \bullet \frac{\pi d}{\pi D} \bullet \frac{d}{D}$$

Because the diameter is equal to twice the radius,

$$\text{gr} \bullet \frac{d}{D} \bullet \frac{2r}{2R} \bullet \frac{r}{R}$$

can be used as well to figure proportion.

## *Using Revolutions per Minute to Determine Torque*

SAMPLE PROBLEM 1

$$\text{Torque}_{WT} \bullet \text{Velocity}_{WT} \bullet \text{Torque}_{Gen} \bullet \text{Velocity}_{Gen}$$

$$\text{Torque}_{Gen} \bullet \frac{\text{Torque}_{WT} \bullet \text{Velocity}_{WT}}{\text{Velocity}_{Gen}} \bullet \frac{(8750 \text{ ft-lbf}) \bullet (22 \text{ rpm})}{1800 \text{ rpm}} \bullet 106.9 \text{ ft-lbf}$$

A simple relationship with torque and velocity between the hub of the wind turbine and generator can be defined as follows:

$$\text{Torque}_{WT} \bullet \text{Velocity}_{WT} \bullet \text{Torque}_{Gen} \bullet \text{Velocity}_{Gen}$$

The gearing ratio is the value at which you change the velocity and torque. Again, it has a very simple equation. The gearing ratio is just a fraction by which you multiple the velocity and torque and can be defined as follows:

$$\text{Gear Ratio} = x\!\!\Big/\!\!y$$

$$\text{Torque}_{Gen} = \text{Torque}_{WT} \bullet \left( x\!\!\Big/\!\!y \right)$$

$$\text{Velocity}_{Gen} = \text{Velocity}_{WT} \bullet \left( x\!\!\Big/\!\!y \right)$$

Suppose a GE 1.5-MW wind turbine generator has nominal speed of 22 rpm and produces a torque of 8,750 ft-lbf. Determine the torque that the generator will experience if the generator has a velocity of 1800 rpm.

Take the same initial conditions and determine torque and velocity for the generator if the gear ratio in the gearbox is 1:115.

## *Gear Failure and Preventive Maintenance*

### Failure Mechanisms

Failures within the gearbox can result from a variety of issues. Increased vibration and noise from the equipment is commonly associated with gear failures, as is wear. Insufficient lubricant-film thickness allows surface-to-surface contact where gear teeth mesh. In optimum operating conditions, gears make contact at points or along lines in small areas or along narrow bands. Each part of the gear tooth surface is in contact only for a short time. Gear tooth surface alignment affects the loading in use, so tooth surface irregularities can be a cause of gear failure over time. Lubrication affects gear teeth as well. If corrosion or debris is present in the lubrication, it will cause wear or other associated failures (see the following list). Finally, temperature can be a factor in gear failure. You can find more about gear failure mechanisms in the "Additional Resources" section of this chapter, but here are some specifics on failure mechanisms you may encounter:

- **Polishing wear:** Polishing wear occurs when slowly rotating gears wear against one another until an almost mirrorlike surface is achieved. Because this occurs at a very slow rate, polishing wear itself is not normally a failure mechanism. However, it can be a catalyst to another type of failure, as the metal particles removed from the gears during the polishing wear could conceivably be deposited in the lubrication system, causing failure. This, of course, can be prevented with regular maintenance of lubrication systems.

- **Moderate to excessive wear:** Moderate wear typically occurs when lubrication-film thickness is insufficient. Like polishing wear, moderate wear is a precursor to failure, as over time, the level of wear may increase to excessive wear. Wear rate can mitigated by increasing oil viscosity, using an extreme-pressure oil, improving the tooth surface finish, or changing the gear geometry to reduce the sliding velocity.

  Excessive wear results in total gear tooth destruction, creating a catastrophic failure that could be due to one or more of the following:

  o The tooth wearing so thin that its bending strength is exceeded

  o Progression of cracks originating at points of tooth surface damage

  o High dynamic loads induced by tooth profile damage

- **Abrasive wear:** Abrasive wear occurs when debris particles are as hard as or harder than the tooth surface and have a diameter greater than the lubrication-film thickness. To prevent this wear, technicians must ensure that the lubrication is clean at all times and must check oil filters regularly.

- **Corrosion wear:** Corrosion of tooth surfaces can result from a breakdown of high-pressure oil as well as from the introduction of outside contaminants. When corrosion occurs on the tooth surface, the finish is damaged, thereby reducing the area of contact. This in turn increases the unit loading on the tooth surface, which leads to accelerated wear.

- **Frosting, scoring, or scuffing:** The terms *frosting*, *scoring*, and *scuffing* are used interchangeably, but all of these types of wear are caused when the asperities of the tooth surface undergo an instantaneous welding, followed by a breaking of the weld. The welding can occur when a critical value is reached, causing a breakdown of the oil film separating the tooth surfaces. This critical value is caused by a combination of load, sliding velocity, and oil temperature. Once the oil film is compromised, metal-to-metal contact occurs. The welding will occur if the surface pressure and sliding velocity are high enough. Typically, the extent to which the welding occurs, combined with the effect of breaking the weld, is the difference between frosting and scoring. High-load, high-speed gears that operate with low-viscosity synthetic lubricants normally create scoring.

  o **Frosting** occurs when only the extreme tips of the asperities are welded and the subsequent break causes little to no damage. The resulting damage is called *frosting* because of the frosted-crystal appearance caused by the micropitting of the surface with no tear marks seen in the sliding direction. During the initial frosting, when the extreme tips of the surface asperities are removed, the contact surface area may be increased, which lowers surface pressure. This lowers the surface pressure, and many times the gears can run for long periods without causing further damage. It is possible to remove the frosting and return the gears to service by polishing the affected area with a very fine grit paper.

  o **Light to moderate scoring** happens when large-scale instantaneous welding of the surface asperities occurs through the combination of temperature, sliding, and load sufficiently above the critical value. When the weld subsequently breaks, the tooth surfaces are scratched and slide on each other. If this condition is not corrected, it will usually progress to complete destruction of the tooth profile. The problem may be corrected by polishing the tooth surface. As is the case with frosting, light to moderate scoring can heal over with continued operation as the tooth surfaces asperities are reduced.

  o **Destructive scoring** takes place when operating conditions are at levels far above the critical point and scoring has progressed from moderate to excessive. Over the long term, destructive scoring will destroy the tooth profile, resulting in breakage as well as metal particle generation that may affect other operations.

  o **Localized scoring** is caused when a nonuniform load is on a tooth surface. Misalignment and local tooth profile errors can both cause localized scoring. Gears with minimal amounts of localized scoring may continue to operate without further damage if the scoring removes the cause of the nonuniform loading (such as a high spot on a profile) and the remaining contact surface is capable of supporting the full load. In some circumstances, initial localized scoring can be indicative of underlying problems, such as misalignment, which could lead to failures of a more catastrophic nature if left uncorrected; in this case, localized scoring can be of diagnostic benefit.

- **Interference** occurs when one tooth surface interferes with another's surface. If the interference can be corrected, in some cases, the damage can be corrected. Tooth interference often indicates poor design or manufacturing.

- **Surface fatigue** happens when the tooth surface is subjected over time to continual application and removal of load. This fatigue usually manifests as pitting and spalling. The extent of the fatigue will depend on the number of load cycles to which the tooth is subjected. A gear designed for long life will not be able to handle surface loads as heavy as those that a gear designed for short life can handle. Even though the fatigue occurs at depths below the tooth's surface, it is called surface fatigue because the tooth's surface is damaged by the fatigue.

- **Fracture** is the most severe form of gear failure. Although the failure mechanisms described here occur over time, a fracture is a sudden breakage that can be cause by a variety of reasons, but it always indicates a need for immediate service to put the turbine back online.

## Avoiding Failures

To avoid failures and service interruptions, it is important for a wind turbine technician to have a maintenance strategy. This will help ensure that the system functions as it is designed to. Corrective action is taken after a fault in the item has been found and is also known as a run-to-failure strategy, but a preventive strategy is most valuable in that it can avoid potential loss time and expensive repairs. The components of a preventive strategy are as follows:

- **Inspection frequency:** The frequency for inspection be based on operating time interval or calendar interval or on some measurable quantity called condition-based monitoring (CBM).

- **Overhaul frequency:** Regularly scheduled maintenance times during which entire sections of turbines are offline for repair and maintenance.

- **Replacement rule for components:** Intervals of part replacement that are based on the manufacturer's recommendations.

- **Management of spare parts:** Keeping an inventory of spare parts available for use as needed.

- **Reliability-centered maintenance:** Reliability-centered maintenance (RCM) operates under the philosophy of "a preventive task is worth doing if it deals directly with preventing a failure in the future."

- **Condition-monitoring system:** For wind turbines, condition-monitoring systems (CMSs) are a new and quickly developing technology. It is quickly becoming commonplace for onshore and offshore wind turbines to be equipped with vibration-based condition monitoring. Moreover, many companies require wind turbines to have some type of installed CMS; otherwise, some components will have a higher replacement frequency in a preventive maintenance program, resulting in higher operating and maintenance costs. CMSs permanently monitor all components of a wind turbine that are subject to wear and failure, such as the gearbox, bearing, and generator.

Studies have shown that significant savings result from having a CMS in place. The general benefits of a CMS are as follows:

- **Avoidance of premature breakdown:** Early fault detection allows prevention of catastrophic failures. For example, late detection of a bearing fault may, in the worst case, mean total destruction of a gearbox.

- **Reduction of maintenance costs:** By using online monitoring, inspections can be avoided or inspection intervals can be increased.

- **Supervision and diagnosis at remote sites:** Large wind turbines are usually built at remote sites. Online monitoring systems can detect any changes at an early stage and, if integrated in a network, can send a warning with diagnostic details to maintenance staff members located elsewhere.

- **Improvement of capacity factor:** A CMS can provide an accurate estimation of the remaining useful life on various mechanisms within the turbine, allowing repairs or replacement actions to be scheduled during time frames with little or no wind. Capacity factor is defined as the ratio between the actual production over a given period of time and the amount of power the turbine would have produced running at full capacity during the same period of time.

For a CMS to work as designed, it must be entirely incorporated into the maintenance strategy. That means that preventive maintenance, or CBM, which is based on the predictive health of the system, is regularly attended to. This is accomplished through periodic inspections; by analyzing offline measurements, oil samples, or SCADA (system control and data acquisition) data; or by continual monitoring of the components during operation.

## Lubrication and Cooling

One of the most important pieces of preventive maintenance is to accurately assess the condition of the lubrication systems on a turbine. The function of the lubrication system is to maintain an oil film on gear teeth and the rolling elements of bearings to minimize surface pitting and wear. Different types of lubricants are available, and the selection of the most suitable depends on gearbox design and its operational condition (location, weather, etc). The quality of the lubrication has been found to be a decisive factor for the service life of the gearbox. Oil temperatures that are too high cause just as much damage as contamination in the oil does. Oil coolers and filters are indispensable for large gearboxes, and so is the careful observance of oil-change intervals. Two alternative methods of lubrication are available:

- **Splash lubrication:** The low-speed gear dips into an oil bath, and the oil thrown up against the inside of the casing is channeled down to the bearings. The advantage of splash lubrication is its simplicity and reliability.

- **Pressure-fed oil:** Oil is circulated by a pump, filtered, and delivered under pressure to the gears and bearings. Pressure-fed lubrication is usually preferred for the following reasons:

  - Oil can be directed where it is required by jets.

- ○ Wear particles can be removed by filtration.

- ○ The oil circulation system enables heat to be removed much more effectively from the gearbox by passing the oil through a cooler.

Lubrication cleanliness is paramount to proper gear function. Contaminants can be generated internally, as debris from tooth wear, or it can be externally introduced, either during routine maintenance or from weather conditions. Studies performed by London's Imperial College have shown that bearing life can be extended by up to seven times with proper lubrication techniques and maintenance. However, with contaminants present in the lubrication, gear damage can result in as little as 30 minutes. Proper lubrication systems and filters, good housekeeping measures during routine maintenance, and temperature control all work together to optimize wind turbine performance.

**Gear Inspection**

Because gear teeth constantly contact one another, causing wear, the wind turbine technician must visually inspect the gears. For most wind energy companies, this maintenance happens semiannually to annually. Although gear maintenance protocol may vary from power company to power company, in general a technician will be responsible for checking gears and lubrication systems.

When you are inspecting gears, you should first look for dirt and/or metal debris on the bottom of the gearbox. This would indicate a potential problem immediately. Next, the technician inspects the gear, using a boroscope, which magnifies images. Look for shiny places, which indicate extensive rubbing, and also for pits, breaks, cracks, and high or low spots.

Next you will look at the lubrication systems of the gearbox. Oil filters and oil levels must be checked and documented. You will be watching for water or dirt and debris in the line, which would indicate a leak. If oil levels are low, this can also indicate a leak. The color and viscosity of the oil is another important factor in determining what maintenance, if any, should occur. If there are any immediate problems, you must return the turbine to proper working order and discover the source of the problem.

Documenting everything you observe and do during routine maintenance is imperative. Whether a wind turbine will be shut down for immediate repairs is a decision an engineer makes on the basis of the findings and recommendations of the turbine technician. Each turbine normally has its own material history book, located on-site, in which a technician can discover whether there is a history of mechanical failure as well as document any potential or current problems. When making recommendations for repair to an engineer, you must also take into account any outside parameters, such as weather and season, because these outside parameters can dictate whether the current conditions are best for maintenance.

SAMPLE PROBLEM 2

Performing maintenance up tower, you have been directed to perform an internal inspection of the gearbox. As you remove the cover, you notice metal shavings at the bottom of the gearbox casing. Explain what actions you would take after this discovery.

## Conclusion

Wind turbine maintenance is often a delicate process that balances intuitive skills, common sense, and gained knowledge. As a wind turbine technician, you must be able to properly analyze the data when checking gears for wear or when assessing the condition of lubricants. At times, you will have to make a decision that is based on prediction, but a wise decision made in a timely manner can prevent more costly repairs in the future. See the "Additional Resources" section at the end of this chapter for other detailed information that will help you make better maintenance assessments on the job.

## References and Additional Resources

### Condition-Monitoring System Information

http://www.nordex-online.com/en/produkte-service/service/condition-monitoring-system.html

http://www.globalspec.com/reference/4998/Wind-Turbine-Technology-Turns-on-Bearings-and-Condition-Monitoring

### Lubrication Information

http://www.machinerylubrication.com/article_detail.asp?articleid=369

http://books.google.com/books?id=g3v3MXzJEZsC&pg=PA798&lpg=PA798&dq=gear+lubrication+in+wind+turbine&source=bl&ots=8hXOAh67y4&sig=QZoSYkJuQ4lv6S420pkxO_ZsIn4&hl=en&ei=aFkkStKwFpGEtwfui8jZBg&sa=X&oi=book_result&ct=result&resnum=2

http://pepei.pennnet.com/display_article/358962/6/ARTCL/none/none/1/Choosing-the-Righ-Wind-Turbine-Lubricant/

### Types of Gears

http://www.gears-manufacturers.com/gear-types/html

http://www.fi.edu/time/Journey/Time/Escapements/geartypes.html

### Gearboxes and Gearbox Fatigue

http://www.windpower.org/en/tour/wtrb/powtrain.htm

http://www.eurekalert.org/features/doe/2009-04/drel-ngs042209.php

### Gear Ratio

http://www.howstuffworks.com/gears.htm

http://www.gears-manufacturers.com/gear-ratios.html

## Review Exercises

1. What is a gear, and how does it function?

2. What is the most common type of gearbox found in a wind turbine?

3. Draw a planetary gearbox.

4. Explain the mechanical advantages of gears in wind energy applications.

5. Define the following terms:
   - Pitch circle
   - Tooth face
   - Circular pitch
   - Backlash

6. List five types of gears and explain how they work.

7. What are the three things you must look at when inspecting the gearbox?

8. Choose and discuss three types of gear failures. How do they happen? What can be done to repair the gears or prevent failure?

9. What are three methods to avoid failures? Describe them.

10. Why is proper lubrication critical to proper function of a wind turbine?

## *Application Exercises*

1. Using the resources at the end of the chapter, as well as at least three other resources from the Internet or industry journals, research wind turbine maintenance. Write an essay analyzing two types of recommended maintenance systems or programs.

2. Using oil samples provided by your instructor, analyze three different stages of oil use on the basis of color and condition. What would your recommendations for replacement and potential ill affects be with each sample?

3. Identify the following types of gears:

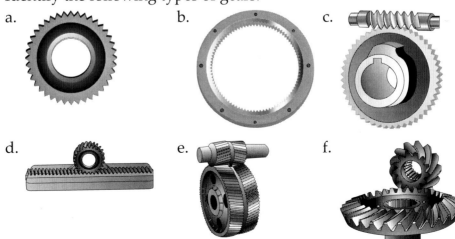

## Chapter Five
# Wind Tower Control Systems

## Learning Objectives

*In the early development of wind energy, the majority of wind turbines were operated at constant speed. Recently, the number of variable-speed wind turbines installed in wind farms has increased, and as demand has increased, more manufacturers now make variable-speed wind turbines. Today's wind turbines are expensive, complex machines that must take advantage of wind flow without sustaining damage from too much wind, lightning, ice, and other weather events. In addition to weather monitoring, temperature control within the tower nacelle is necessary for all mechanical and electrical systems to operate as designed (which in turn keeps the wind tower functioning appropriately), and all of these systems must be controlled by a centralized computer. These variable-speed turbines are controlled by an electrical, a hydraulic, or a mechanical system. In the United States, most turbines, but not all, are electrically controlled.*

*In this chapter, you will learn about:*

- *The three power-control systems for wind turbines*
- *The electrical control system components, such as circuit-protection devices, sensors, relays, contactors, actuators, timers, counters, motors, and various types of DC and AC drives*
- *The application of electrical systems learned to laboratory practicals*
- *Temperature-monitoring systems*
- *Vibration-monitoring systems*

## *Introduction*

Wind turbines are designed to produce electrical energy as cheaply and efficiently as possible. They are therefore generally designed so that maximum yield occurs at wind speeds of about 33 mph (primarily depending on the size of the turbine).

In stronger winds, unless a part of the excess wind energy is eliminated, the turbine can be damaged. Because of this, all wind turbines are designed with some sort of power control. There are three different power-control systems on utility-scale modern wind turbines: pitch-controlled, passive stall-controlled, and active stall-controlled.

# Types of Turbine Power-Control Systems

### Pitch-Controlled Wind Turbines

On a pitch-controlled wind turbine, the turbine's electronic controller checks the power output of the turbine constantly. If the power output becomes too high, it sends an order to the blade pitch mechanism, which immediately pitches, or turns, the rotor blades slightly out of the wind to slow rotation. When the wind speed decreases, the blades are turned back into the wind. The rotor blades have to be able to turn around their longitudinal axis.

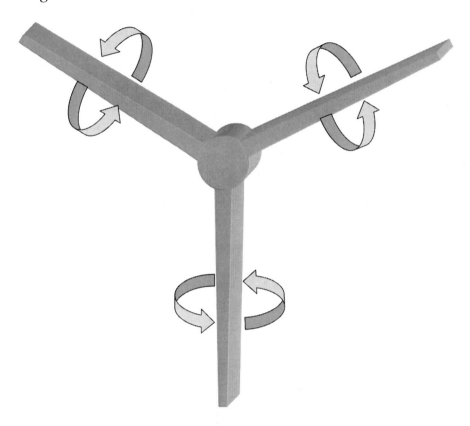

During strong winds, this system not only prevents damage to the turbines but also maximizes their power.

In normal operating conditions, the blades constantly change pitch, a fraction of a degree at a time while the rotor is turning, to adjust to changing wind conditions (direction, speed, etc.). This changing pitch is achieved through an integrated system that uses anemometers and wind vanes to relay changing wind conditions to a computer system, called an up-tower controller, located in the nacelle. An anemometer is an electronic device, located on top of the nacelle, that is used for measuring wind speed. A weather vane detects wind direction. The information from these devices is relayed to the up-tower controller. In turn, the up-tower controller relays the information to the hydraulic power plant located within the nacelle, which releases oil at an appropriate pressure to rotate the rotor blades to the optimum angle to maximize output for all wind speeds. Smaller turbines may use a mechanical-type pitch control rather a hydraulic system.

**Passive Stall-Controlled Wind Turbines**

Currently, approximately two thirds of the wind turbines in the world are stall-controlled machines. Stall-controlled wind turbines (also known as passive stall-controlled turbines) differ from pitch-controlled wind turbines in that the blades are bolted directly to the hub at a fixed angle. The primary differences in this design are the geometry of the blade and the reaction of the blade to increasing wind speed. If the wind speed becomes too high, turbulence is created on the trailing edge of the blade (the side not facing the wind), causing the blade to stall. The stall occurs because of a lack of lifting force acting on the rotor.

Advantages of the stall-controlled system are that moving parts in the rotor itself are avoided and a complex control system is not necessary. However, stall control involves a very complex aerodynamic design and related design challenges in the structural dynamics of the whole wind turbine, for example, to avoid stall-induced vibrations.

If you look closely at a rotor blade for a stall-controlled wind turbine, you will notice that the blade is slightly twisted along its longitudinal axis. This is done in part to ensure that the rotor blade stalls gradually, rather than abruptly, when the wind speed reaches its critical value.

**Active Stall-Controlled Wind Turbines**

On an active stall-controlled turbine, the rotor blades can be pitched, as with a pitch-controlled wind turbine. The difference between the two is that when the machine reaches its rated power, the blades of an active stall-controlled turbine will pitch in the opposite direction, increasing their angle to the wind and going into a deeper stall. Thus, the excess energy in the wind can be avoided.

An advantage of the active stall-controlled machine is that the power output can be more accurately controlled than in the passive stall-controlled turbine, resulting in no generator strain during gusty winds.

In general, turbines have the capability of running at rated power in all high wind speeds. This is not the case, however, with the active stall-controlled turbines, which usually have a drop in the electrical power output for higher wind speeds, as the rotor blades go into deeper stall.

The active stall-control system is often installed in large turbines (1 MW and more).

## Control of Electronic Circuits

As you have seen, wind turbines use one of three control systems. In the United States, the majority of turbines use electromechanical systems, as opposed to hydraulic or mechanical systems, to manipulate these control systems. To keep the turbines reliable and simple, electromechanical relays are used to send electrical current to the receiving device that controls the stall. This section introduces you to the various components used to control the turbine and its operation.

## Electromechanical Relays

An electromechanical relay is a device that either completes or interrupts a circuit by causing physical contact between two electrical points. Two circuits make up an electromechanical relay: an energizing circuit and a contact circuit. When a relay coil is energized, current flows through the coil, creating a magnetic field. If the relay is a DC unit, the polarity is fixed, but if it is an AC unit, the current alternates polarities 120 times per second. Regardless of whether it is AC or DC, the relay functions in the same way, attracting a ferrous plate that is part of the armature to the magnetic coil. One end of the armature pivots, whereas the other end opens and closes the contacts. Although this explanation is simplistic, relays are designed, engineered, and built for their specification application from light duty to heavy duty. All relays work off Kirchhoff's laws, two electrical engineering precepts named for Gustav Kirchhoff, that prescribe the electrical function of the relays in this manner:

Kirchhoff's circuit law (KCL): The sum of the current entering a node equals the sum of the current exiting a node.

$$\sum I_{in} - \sum I_{out}$$

Kirchhoff's voltage law (KVL): The sum of voltages around a loop equal zero.

$$\sum V_{Loop} - 0$$

## Relay Construction

A relay is a simple electromechanical switch made up of an electromagnet and a set of contacts. The relay is constructed of four primary parts: armature, electromagnet, spring, and contacts.

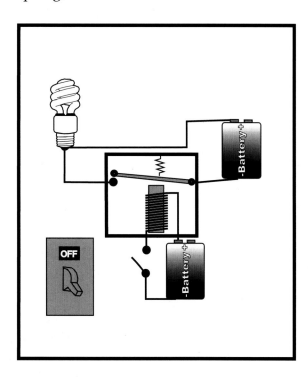
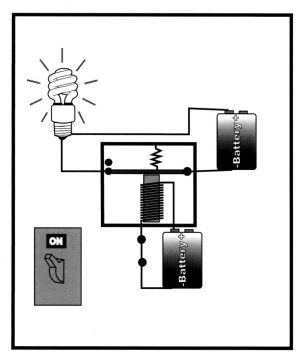

## Contactors

When a relay is used to switch a large amount of electrical power through its contacts, it is designated by a special name: contactor. Contactors typically have multiple contacts, and those contacts are usually (but not always) oriented as normally open, so that power to the load is shut off when the coil is de-energized. Perhaps the most common industrial use for contactors is the control of electric motors.

There are three major types of AC controllers in use today: low-voltage protection (LVP), low-voltage release (LVR), and low-voltage release effect (LVRE) controllers.

- **LVP motor controller:** The main purpose of an LVP controller is to de-energize the motor in a low-voltage condition and keep it from restarting automatically on return of normal voltage.

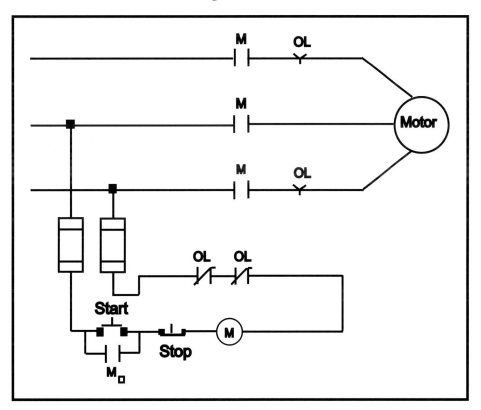

Basic operation: Pressing the Start button energizes the contactor coil (M), which closes the M and $M_o$ contacts. The Start button is spring loaded; when it is released, the contactor coil remains energized, which keeps the M and $M_o$ contacts closed, creating a path for current flow to the motor.

If a low-voltage condition exists, the contactor coil (M) will no longer be energized, causing the M and $M_o$ contactors to open, thus de-energizing the motor.

Depressing the Stop button de-energizes the contactor coil (M), thus opening the M and $M_o$ contacts and stopping the motor.

- **LVR motor controller:** The purpose of the LVR controller is to de-energize the motor in a low-voltage condition and restart the motor when normal voltage is restored. This type of controller is used primarily on small and/or vital loads.

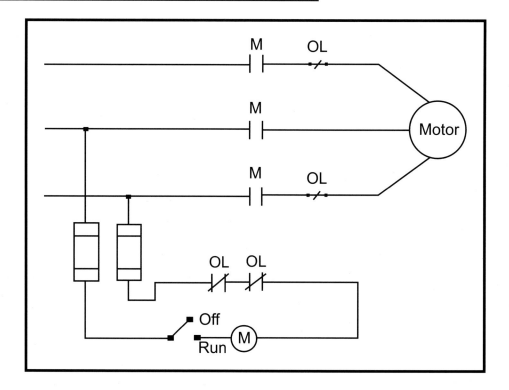

Basic operation: Place the Start switch in the Run position, which energizes the contactor coil (M), closing the M contacts and starting the motor.

When a low-voltage condition occurs, the contactor coil (M) drops out, opening the M contacts and de-energizing the motor. When normal voltage is restored, the M coil is again energized, closing the M contacts and restarting the motor.

- **LVRE motor controller:** This controller maintains the motor across the line at all times. This type of controller is manual and is found mostly on small loads that must start automatically on restoration of voltage. An LVRE controller may or may not contain overloads. If overloads are used, they will be placed in the lines to the load.

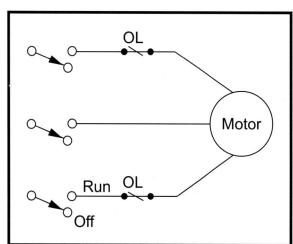

## Time-Delay Relays

A time-delay relay is a relay that stays on for a certain amount of time once activated. The time-delay relay is made up of a simple adjustable timer circuit, which controls the actual relay. Some relays are constructed with a kind of "shock absorber" mechanism attached to the armature, which prevents immediate, full motion when the coil is either energized or de-energized. This addition gives the relay the property of time-delay actuation. Time-delay relays can be constructed to delay armature motion on coil energization, de-energization, or both.

Time-delay relay contacts must be specified not only as either normally open or normally closed but also regarding whether the delay operates in the direction of closing or in the direction of opening. The following is a description of the four basic types of time-delay relay contacts:

- **Normally open, timed closed:** The NOTC contact, or normally open, timed closed contact, is customarily open when the coil is de-energized. The contact is closed when power is applied to the relay coil, but only after the coil has been continuously powered for the specified amount of time. In other words, the direction of the contact's motion (either to close or to open) is identical to a regular NO (normally open) contact, but there is a delay in closing direction. Because the delay occurs in the direction of the coil being energized, this type of contact is alternatively known as a normally open, on-delay. In the schematic representation of a NOTC relay, you can see a graphical representation of coil power and contact position relative to the coil power. In this instance, the contact position closes 10 seconds after coil power is energized and opens when coil power is de-energized.

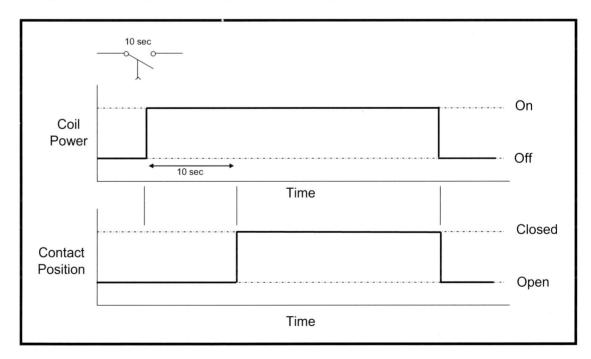

- **Normally opened, timed open:** of the NOTO contact, or normally opened, timed open contact, is normally open when the coil is de-energized and is closed by applying power to the relay coil. However, unlike what happens with the NOTC contact, the timing action of the NOTO contact occurs on the coil's being de-energized rather than energized. Because the delay occurs in the direction of coil de-energization, this type of contact is alternatively known as a normally open, off-delay. In the schematic representation of a NOTO relay, you can see a graphical representation of coil power and contact position relative to the coil power. In this instance, the contact position closes 10 seconds after coil power is energized and opens when coil power is de-energized.

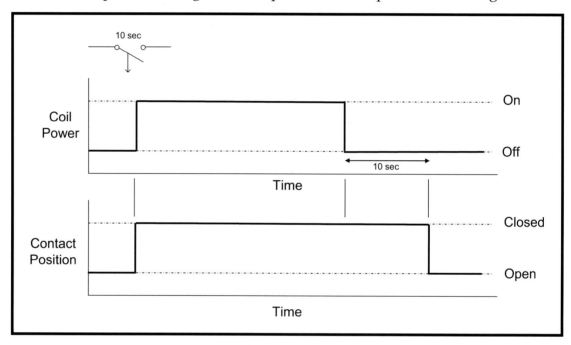

- **Normally closed, timed open:** The NCTO contact, or normally closed, timed open contact, is normally closed when the coil is de-energized. The contact is opened with the application of power to the relay coil, but only after the coil has been continuously powered for the specified amount of time. In other words, the direction of the contact's motion (either to close or to open) is identical to a regular NC (normally closed) contact, but there is a delay in the opening direction. Because the delay occurs in the direction of coil energization, this type of contact is alternatively known as a normally closed, on-delay. In the schematic representation of an NCTO relay, you can see a graphical representation of coil power and contact position relative to the coil power. In this instance, the contact position closes 10 seconds after coil power is energized and opens when coil power is de-energized.

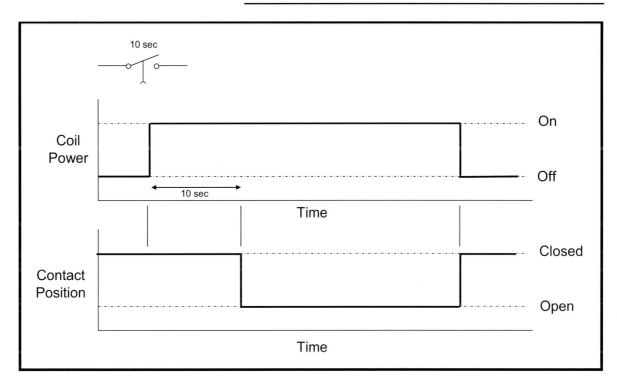

- Normally closed, timed closed: The NTCT contact, or normally closed, timed open contact, is normally closed when the coil is de-energized and is opened by applying power to the relay coil. However, unlike the NCTO contact, the timing action occurs on de-energization of the coil rather than on energization. Because the delay occurs in the direction of coil de-energization, this type of contact is alternatively known as normally closed, off-delay.

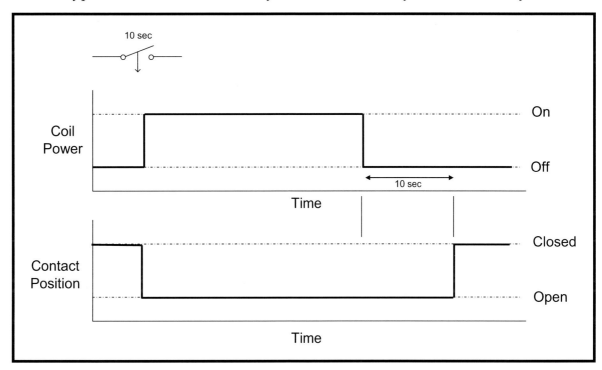

## Protective Relays

The function of protective relaying is to cause the prompt removal from service of any element of a power system when it experiences a short circuit or when it starts to operate in any abnormal manner that might cause damage or otherwise interfere with the effective operation of the rest of the system. The relaying equipment is aided in this task by circuit breakers that are capable of disconnecting the faulty element when they are called on to do so by the relaying equipment.

Circuit breakers are generally located so that each generator, transformer, bus, transmission line, and so on, can be completely disconnected from the rest of the system. These circuit breakers must have sufficient capacity so that they can momentarily carry the maximum short-circuit current that can flow through them and then interrupt this current. They must also withstand closing in on such a short circuit and then interrupting it according to certain prescribed standards.

If protective relays and circuit breakers are not economically justified, fusing is employed. Although the principal function of protective relaying is to mitigate the effects of short circuits, other abnormal operating conditions arise that also require the services of protective relaying. This is particularly true of generators and motors.

## Solid-State Relays

One drawback of an electromechanical relay is the use of moving parts. A solid-state relay (SSR) is an electronic switch that has no moving parts. SSRs are classified by the nature of the input circuit. You may encounter two types of SSRs in a wind turbine:

- **Photo-coupled SSR:** This type of SSR is controlled by a low-voltage signal that is isolated optically from the load. The control signal in a photo-coupled SSR typically energizes an LED, which activates a photo-sensitive diode. The diode turns on a back-to-back thyristor, a silicon-controlled rectifier, to switch the load.

- **Transformer-coupled SSR:** In this type of SSR, the control signal is applied, either through a DC/AC converter for DC current or directly when the current is AC, to the primary voltage of a small, low-power transformer. The resulting secondary voltage is used to trigger the thyristor switch.

SSRs have advantages and disadvantages, which must be weighed before a decision on whether to use them in a given application is made. Wikipedia (http://en.wikipedia.org/wiki/Solid-state_relay) lists the following pros and cons for SSRs:

ADVANTAGES

- SSRs are faster than electromechanical relays; their switching time is dependent on the time needed to power the LED (light-emitting diode) on and off, on the order of microseconds to milliseconds.

- They have increased lifetime because there are no moving parts, and thus no wear.

- They produce clean, bounceless operation.

- They produce less electrical noise when switching.
- They can be used in explosive environments where a spark must not be generated during turn-on.
- They operate in total silence.
- They are smaller than a corresponding mechanical relay.

**DISADVANTAGES**

- SSRs fail or short more easily than electromechanical relays do.
- They produce increased electrical noise when conducting.
- They produce higher impedance when closed, meaning they have increased heat production.
- They have lower impedance when open.
- They have reverse leakage current when open (µA range).
- They carry the possibility of false switching due to voltage transience.
- They are often more expensive than comparable electromechanical relays.
- They require an isolated bias supply for gate charge circuit.
- They have a higher transient reverse recovery time because of the presence of a body diode.

## Other Types of Switches

In addition to relays, you may encounter proximity switches and/or limiter switches. Like relays, both of these switches signal an electronic circuit when to open or close.

**PROXIMITY SWITCHES**

Proximity switches are used to open or close a circuit when they come within a specific distance of an object. There are four common types:

- **Capacitive proximity switches** sense distance to objects by detecting a change in capacitance. The sensor has a metal plate and a radio frequency oscillator connected to it. When an object comes to a certain distance to the sensor, the radio frequency changes and sends a signal to the switch to open or close it.
- **Inductive proximity switches** sense distance to objects by generating magnetic fields. A coil of wire has electricity running through it and measures current. If a part made of metal comes within a close distance, the current will increase and the switch will open or close.
- **Acoustic proximity switches** use sound to send a signal to a circuit. The time necessary for the signal to return is used to trip the transducer and open or close a switch.
- **Infrared proximity switches** work by sending out beams of IR light. A photodetector on the switch detects reflections of the light. As the light source is interrupted by a nearby object, the switch will be triggered to open or close.

**LIMIT SWITCHES**

A limit switch is an electromechanical device that consists of an actuator mechanically linked to a set of contacts. When an object comes into contact with the actuator, the device operates the contacts to make or break an electrical connection. Limit switches are used in a variety of applications and environments because of their ruggedness, simple visible operation, ease of installation, and reliability.

# Common Control System Components in Wind Turbines

The control system in a wind turbine consists of a central SCADA (system control and data acquisition) computer, which functions as the nerve center for the wind tower. It connects each turbine to the meteorological station and grid station through a remote interface unit (RIU), using a site network as shown in the system schematic in this chapter. For some projects, where only a central connection to an existing SCADA system is available, an RIU server is provided, which runs several virtual RIUs. Systems may also have a mixture of real and virtual RIUs. The system is expandable so that several SCADA computers can be networked together for large wind farms. There is no limit to the number of turbines that can be monitored with the system.

There are several advantages to this system:

- It is completely independent, traceable, and transparent in operation. The users specify what they want to see, control, record, and report.
- The RIUs for turbine, grid, and meteorological station are independent, providing maximum data integrity.
- Remote access is easy, requiring only a Web browser. No additional software or hardware is required to access the system.
- It is easily customized to meet the different needs of turbine manufacturers, wind farm developers, wind farm operators, wind farm owners, and financiers.
- It operates seamlessly with machines from any wind turbine manufacturer.
- It has a common graphical interface, allowing users to have a consistent front-end display for all their wind farms.
- It uses a common reporting system, enabling users to produce reports with the same format for all their wind farms.
- It uses a common format for the database, allowing the same data processing to be used for wind farms with different makes of turbines.
- It allows comparison of actual to expected energy and revenue output using a wind farm power curve.
- It has grid and substation interface units for control and monitoring of the wind farm electrical system.
- It has meteorological station interface units for independent wind speed and meteorological parameter monitoring and analysis.

This is a schematic of a typical SCADA system. The details of any installation will be project-specific.

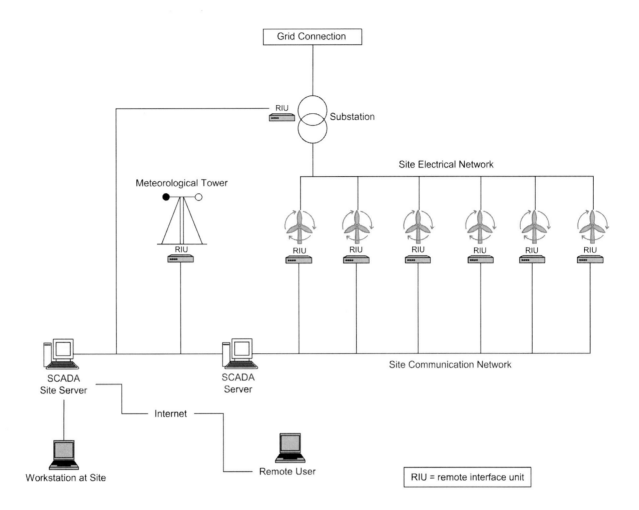

### Remote Interface Units

A key feature of the system is the RIU at each turbine, meteorological station, and grid station. In the turbines, these units are connected between the turbine controller and the site communications network. They have local processing and storage and provide the following benefits:

- The data sample rate is independent of the site communications network and depends only on the speed of communication of the turbine controller. The higher sampling rate improves the accuracy of the summary statistics that are produced.

- All sampled data are stored in high-resolution data files. These are stored on the RIU until they can be transferred to the SCADA server. There is continuous data coverage; plus, additional *trip files* are created when particular events occur.

- If available, turbine controller trip files are retrieved from the turbine controller and stored on the RIU until they can be transferred to the SCADA server.

- Processed data is stored locally in a queue so that no data are lost if the site communications network is temporarily unavailable. When the network is available, the queue is downloaded to the SCADA server.

- The RIU provides a standard interface between the SCADA system and the turbine controller, allowing different turbine types to be accommodated without requirement for system modifications.

- The RIU can provide additional functionality that is not included in the standard turbine controllers.

- The RIUs can be expanded to provide additional input and output so that additional sensors can be connected to the system. This may include noise-monitoring stations and safety and access control equipment.

- Software updates are automatically downloaded to the RIUs from the SCADA server.

- Configuration information is automatically downloaded to the RIUs from the SCADA server.

- The RIU clock is automatically corrected to the central SCADA server clock.

### Programmable Logic Controllers

Programmable logic controllers (PLCs) are another type of digital computer system used to control electromechanical processes, such as blade pitch and turbine yaw. The PLC differs from a general-purpose computer in that it is designed with a multiple inputs/output arrangement. It is compatible with extended temperature ranges, is immune to electrical interference, and resists impact and vibrations. The programs that control machine operation are typically stored in battery-backed or nonvolatile memory. A PLC is an example of a real-time system because output results must be produced in response to input conditions within a bounded time; otherwise, unintended operation will result.

PLCs monitor data inputs, make decisions, and control outputs to automate machines and processes, as shown in the following illustrations.

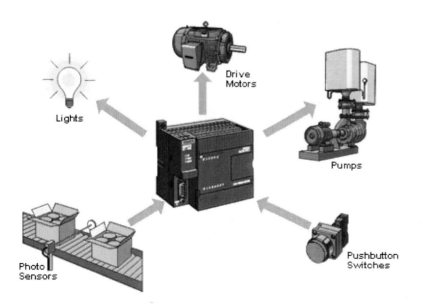

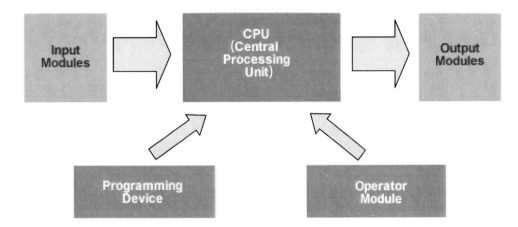

In the diagram of integration of a PLC into a system, the Start/Stop button can be used as a sensor, which sends a signal to the PLC. The PLC's output is connected to a motor starter that is used to start and stop the motor.

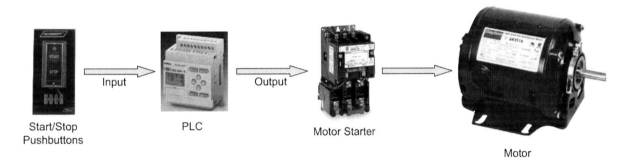

There are several advantages to the use of PLCs, including

- Their smaller size today compared with that of early-model PLCs
- Easier and faster system changes than with older wired relays
- Centrally available diagnostics
- Immediate documentation
- Faster and less costly duplication of applications than with older wired relays

### Temperature Measurement

As you will learn in Chapter Six, "Wind Turbine Materials and Failure Systems," heat is a common failure mechanism in wind towers. Therefore, towers are equipped with a variety of temperature-measurement capabilities to optimize performance:

- **Thermocouples:** Thermocouples consist essentially of two strips or wires made of different metals and joined at one end. Changes in the temperature at that juncture induce a change in electromotive force (EMF) between the other ends. As the temperature goes up, this output EMF of the thermocouple rises, though not necessarily linearly. Some of the advantages of thermocouples are
  - Low cost
  - Lack of moving parts, which means that they are less likely to be broken

○ Wide temperature range

○ Reasonably short response time

○ Reasonable repeatability and accuracy

A few of their disadvantages are that

○ Their sensitivity is low and their low-voltage output may be masked by noise. This problem can be decreased, but not eliminated, by better signal filtering, shielding, and analog-to-digital conversion.

○ Their accuracy, usually no better than 0.5°C (0.9°F), may not be high enough for some applications

○ Traditional thermocouples require a known temperature reference, usually 0°C (32°F) ice water. Modern thermocouples, however, rely on an electrically generated reference.

○ Their nonlinearity can be bothersome. Fortunately, detail calibration curves for each wire material can usually be obtained from vendors.

• **Infrared radiators:** Infrared (IR) sensors are noncontacting devices. They infer temperature by measuring the thermal radiation emitted by a material. IR radiation is part of the electromagnetic spectrum, which includes radio waves, microwaves, visible light, and ultraviolet light, as well as gamma rays and X-rays. The IR range falls between the visible portion of the spectrum and radio waves. IR wavelengths are usually expressed in microns,

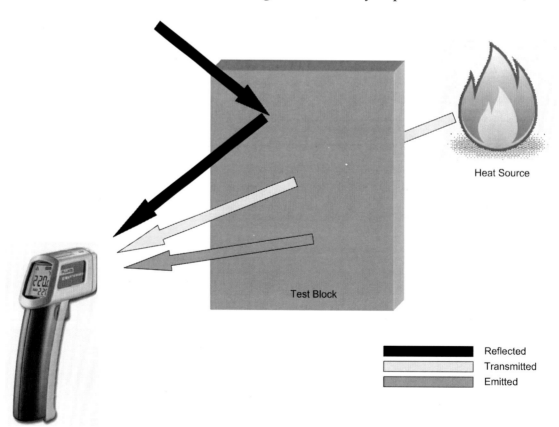

*Example of reflection, transmission and emission of energy on an object*

with the spectrum extending from 0.7 to 1000 μ. Only the s0.7- to 14-μ band is used for IR temperature measurement.

Using advanced optic systems and detectors, noncontact IR thermometers can focus on nearly any portion or portions of the 0.7- to 14-μ band. Because every object (with the exception of a blackbody) emits an optimum amount of IR energy at a specific point along the IR band, each process may require unique sensor models with specific optics and detector types.

An object reflects, transmits, and emits energy, as shown in the diagram.

The intensity of an object's emitted IR energy increases or decreases in proportion to its temperature. It is the emitted energy, measured as the target's emissivity, that indicates an object's temperature. Emissivity is a term used to quantify the energy-emitting characteristics of different materials and surfaces. IR sensors have adjustable emissivity settings, usually from 0.1 to 1.0, which allow accurate temperature measurements of several surface types. The emitted energy comes from an object and reaches the IR sensor through its optical system, which focuses the energy onto one or more photosensitive detectors. The detector then converts the IR energy into an electrical signal, which is in turn converted into a temperature value that is based on the sensor's calibration equation and the target's emissivity. This temperature value can be displayed on the sensor or, in the case of the smart sensor, converted to a digital output and displayed on a computer terminal.

- **Resistance temperature detectors (or thermistors):** Resistance temperature detectors (RTDs) capitalize on the fact that the electrical resistance of a material changes as its temperature changes. Two key types are the metallic devices (commonly referred to as RTDs) and thermistors. As their name indicates, RTDs rely on resistance change in a metal, with the resistance rising more or less linearly with temperature. Thermistors are based on resistance change in a ceramic semiconductor; the resistance drops nonlinearly with temperature rise.

  The RTD is one of the most accurate temperature sensors. Not only does it provide accuracy but it also provides stability and repeatability. RTDs are also relatively immune to electrical noise and therefore are well suited for temperature measurement in industrial environments, especially around motors, generators, and other high-voltage equipment.

- **Bimetallic sensors:** Bimetallic devices take advantage of the difference in rate of thermal expansion between different metals. Strips of two metals are bonded together. When they are heated, one side will expand more than the other, and the resulting bending is translated into a temperature reading by mechanical linkage to a pointer. These devices are portable, and they do not require a power supply, but they are usually not as accurate as thermocouples or RTDs, and they do not readily lend themselves to temperature recording, as is shown in the diagram.

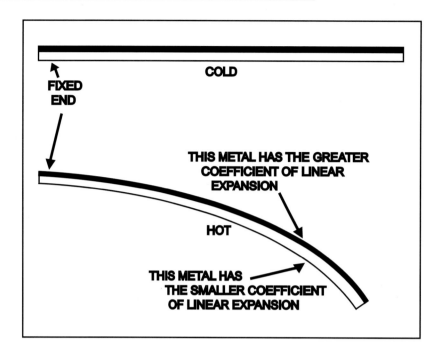

- **Liquid expansion devices:** Fluid-expansion devices, typified by the household thermometer, generally come in two main classifications: the mercury type and the organic-liquid type. Versions employing gas instead of liquid are also available. Mercury is considered an environmental hazard, so there are regulations governing the shipment of devices that contain it. Fluid-expansion sensors do not require electric power, do not pose explosion hazards, and are stable even after repeated cycling. However, they do not generate data that is easily recorded or transmitted, and they cannot make spot or point measurements.

- **Change-of-state devices:** Change-of-state temperature sensors consist of labels, pellets, crayons, lacquers, or liquid crystals whose appearance changes once a certain temperature is reached. They are used, for instance, with steam traps. When a trap exceeds a certain temperature, a white dot on a sensor label attached to the trap will turn black. Response time is typically measured in minutes, so these devices often do not respond to transient temperature changes, and accuracy is lower than with other types of sensors. Furthermore, the change in state is irreversible, except in the case of liquid-crystal displays. Even so, change-of-state sensors can be handy when you need confirmation that the temperature of a piece of equipment or a material has not exceeded a certain level, for instance for technical or legal reasons during product shipment.

### Vibration-Monitoring Control Systems

Two drivers for the rapid growth in wind-generation capacities are the increasing size of wind turbine generators and the declining cost due to economies of scale and design and manufacturing experience. Wind facilities being developed today commonly use machines rated at 1.5 MW or greater. This concentration of production capacity in fewer units with higher capacity potentially reduces the operations and maintenance effort for the entire installation. At the same time, the costs associated with any one failure are increased. The risk associated with a critical component failure also increases with the immaturity of the machine design.

Fortunately, condition-monitoring technology developed for the allied electric power production, marine propulsion, and process industries is available and is beginning to be applied to wind turbines. Cost-effective and robust online monitoring and predictive maintenance technology can detect impending problems and allow repairs to be scheduled during low wind periods, thus avoiding both catastrophic losses and significant repair expense.

Vibration analysis is the most known technology applied for condition monitoring, especially for rotating equipment. The type of sensor used depends more or less on the frequency range, relevant for the monitoring:

- Position transducers for the low-frequency range
- Velocity sensors in the middle-frequency area
- Accelerometers in the high-frequency range
- SEE sensors (spectral emitted energy) for very high frequencies (acoustic vibrations)

Examples can be found for safeguarding of

- Shafts
- Bearings
- Gearboxes
- Compressors
- Motors
- Turbines (gas and steam)
- Pumps

For wind turbines, this type of monitoring is applicable for monitoring the wheels and bearings of the gearbox, bearings of the generator, and the main bearing.

Signal analysis requires specialized knowledge. Suppliers of the system offer turnkey systems, which include signal analysis and diagnostics. The monitoring itself is also often executed by specialized suppliers who maintain the components. The costs are compensated for by reduction of production losses, as wind-induced vibration poses a strong threat to wind towers. Application of vibration-monitoring techniques and working methods for wind turbines differ from other applications with respect to

- The dynamic load characteristics and low rotational speeds. In other applications, loads and speed are often constant during longer periods, which simplify the signal analysis. For more dynamic applications, such as wind turbines, the experience is very limited.

- The high investment costs in relation to costs of production losses. The investments in conditions-monitoring equipment is normally covered by reduced production losses. For wind turbines, especially for land applications, the production losses are relatively low. Therefore, the investment costs should be offset by reduction of maintenance cost and reduced damage costs.

## Conclusion

Many of today's wind turbines are variable speed rather than constant speed and thus are complex systems. In this chapter, you learned about the power-control systems, electrical control system components, and temperature- and vibration-monitoring systems used in modern turbines. Knowing how these items operate will help you be able to properly maintain turbines.

## References and Additional Resources

### Turbine Control Systems
http://zone.ni.com/devzone/cda/tut/p/id/8189

### Electrical Relays
http://electronics.howstuffworks.com/relay1.htm

### Time-Delay Relays
http://www.aaroncake.net/circuits/relaytim.asp

http://www.allaboutcircuits.com/vol_4/chpt_5/3.html

### Solid-State Relays
www.omega.com/temperature/Z/pdf/z124-127.pdf

http://en.wikipedia.org/wiki/Solid_state_relay

### Turbine Control Systems
http://www.wind-energy-the-facts.org/en/part-i-technology/chapter-4-wind-farm-design/infrastructure/scada-and-instruments.html

### Temperature-Measurement Systems
http://www.facstaff.bucknell.edu/mastascu/elessonsHTML/Sensors/TempR.html

http://www.engineeringtoolbox.com/temperature-measurement-t_50.html

## Review Exercises

1. What are the three major types of wind turbine control systems used?

2. What is an electromechanical relay?

3. Describe the three types of AC controllers used in the United States today.

4. How does a time-delay relay work?

5. List and describe the differences between the four major types of time-delay relays.

6. List the types of proximity switches and describe them.

7. What does a SCADA computer do? Why is this important in wind energy?

8. Describe two types of temperature measurements used in wind towers. What do you think the advantages and disadvantages of each are?

9. Why is vibration a danger to wind towers?

10. What kind of systems detect vibration?

## *Application Exercises*

1. Use the Internet to research vibration-monitoring control systems. Choose two of the systems you learn about and write an analysis of your findings Include the pros and cons of each system.

2. A hydraulic cylinder in the hub rotates the blades on the basis of various parameters such as wind speed and direction. Once the circuitry discussed in this chapter has done its job, the rest is up to Pascal's law.

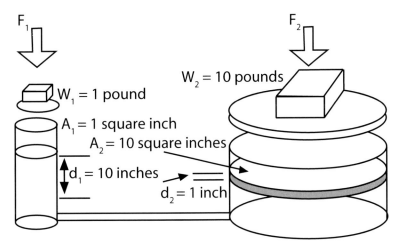

The formulas that relate to this are shown below:

$$P_1 = P_2$$

(because the pressures are equal throughout).

Because pressure equals force per unit area, then it follows that

$$\frac{F_1}{A_1} = \frac{F_2}{A_2}$$

It can be shown by substitution that the values shown above are correct:

$$\frac{1 \text{ lbf}}{1 \text{ in}^2} = \frac{10 \text{ lbf}}{10 \text{ in}^2}$$

Because the volume of fluid pushed down on the left side equals the volume of fluid that is lifted up on the right side, the following formula is also true:

$$V_1 = V_2$$

By substitution,

$$A_1 d_1 = A_2 d_2$$

$A$ = cross-sectional area

$d$ = the distance moved

*or*

$$\frac{A_1}{A_2} = \frac{d_2}{d_1}$$

Given these simple formulas, try to answer the following questions:

A hydraulic press has an input cylinder that has a diameter of 1 in and an output cylinder with a diameter of 6 in.

a.  Find the force exerted by the output piston when a force of 10 pounds (10 lbf) is applied to the input piston.

Given:

$F_1 = 10$ lbf

$D_1 = 1$ in

$D_2 = 6$ in

b.  If the input piston is moved through 4 inches, how far is the output piston moved?

Given:

$d_1 = 4$ in

$A_1 = 0.785$ in²

$A_2 = 28.274$ in²

3.  The output shaft of a hydraulic cylinder that rotates the blade needs to move 10 in to rotate the blade 90 degrees. Knowing the diameter of the output piston is 2.25 in and the diameter of the input piston is 5 in, determine the

length that the input piston needs to move and the input force required if the output force is 5 lbf.

Given:

$d_1 = 2$ in

$d_2 = 10$ in

$D_1 = 5$ in

$D_2 = 2.25$ in

$F_2 = 5$ lbf

*Chapter Six*

# Wind Turbine Materials and Failure Systems

## Learning Objectives

*As technology has continued to develop, resulting design improvements have been adapted to newer generations of wind turbines. Advances in materials science, computer science, aerodynamics, testing, and power electronics have all been adapted for the modern wind turbine. Therefore, you must understand the materials and design considerations used in the manufacturing of wind turbines.*

*In this chapter, you will learn about:*

- *The materials used to build a wind turbine*
- *The design characteristics of the wind turbine blade*
- *The common manufacturing process for wind turbine blades*
- *The impact of heat generation on various materials and heat-control mechanisms*

## *Introduction*

Progress is swift. As recent as the late 1980s, the largest wind turbines had blades with a diameter of 75 feet and were capable of producing only 50 kW of electricity. Today, the largest wind turbine in the world has a capacity of 7 MW and can produce enough electricity to supply more than 1500 households—on a single turbine! With a rotor diameter of 413 feet and a sweep the near equivalent of two football fields, the today's wind turbines are impressive indeed.

Since the mid-1980s, the size of wind turbines has continued to increase. For example, the size for machines on land has gone from 50 kW to 3 MW, and machines up to 7 MW are now planned for the ocean. According to Andy Swift, ScD, PE, director of the Wind Science and Engineering Research Center at Texas Tech University, designs as large 10 MW are being considered for development.

Which size makes most sense? Larger or smaller? Smaller turbines will require more land and a greater number of turbines to produce the same amount of electricity, whereas larger turbines will provide more capacity without burdening the land space yet their production will require the innovation of ever-lighter materials.

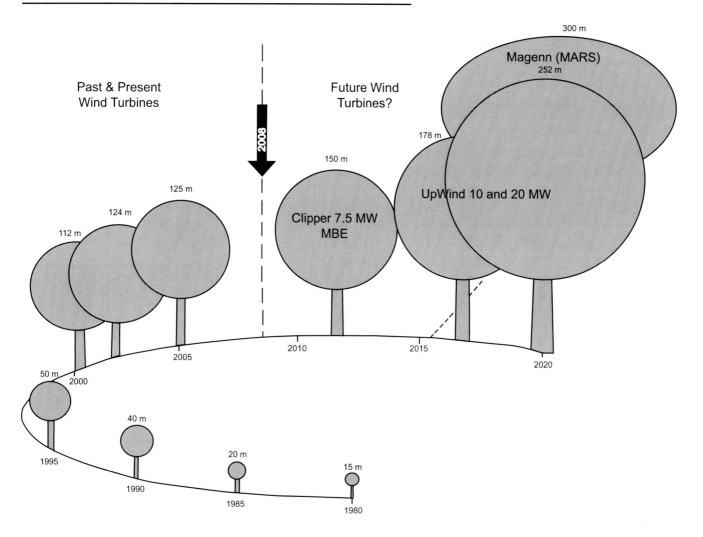

## Design Considerations

New design standards have improved the reliability and performance of current wind turbines. Because of new designs, wind energy has continued to become more cost-effective, so much so that even without incentives such as production tax credits for the energy companies, there will be continued growth in the industry. In areas where the government offers these incentives, there continues to be even more rapid growth in this industry as well as related fields.

Consider the area of computer science, for example. This discipline helps not only with control but also with the design process. Testing methods include many never-before-available sensors, data-collection systems, and analysis equipment, which allow designers the ability to test new designs. Power electronics continue to be applied to modern designs. They allow a wind turbine's generator the ability to provide smooth power to the grid and allow the use of asynchronous generators, which have the ability to produce power at varying rates of speed. GE's new Mark 6 design uses solid-state control systems that eliminate mechanical relays and wiring, thus making the system faster to repair.

In the United States, we have continued to see large-scale wind farms being developed. In Europe, large-scale farms are not generally as prevalent. Regardless of which continent wind turbines are destined for, solid and innovative design principles are key to a turbine's efficient operation. The following is a list of some important design variables for a wind turbine:

- **Rotor diameter:** A larger rotor captures more energy but costs more.

- **Generator capacity:** A larger generator captures more energy at higher wind speeds but also costs more. The rotor diameter and generator capacity must match. The optimal match is dependent on the site's wind conditions.

- **Hub height:** Wind speeds increase with hub height, but increased height means a greater cost to manufacture the tower.

- **Rotor blade design:** The blades have a slight twist, which can be optimized to capture the maximum amount of wind power.

- **Power control:** Should a turbine be active pitch or passive stall? An active pitch-controlled system allows the pitch to be continuously optimized for maximum power. Active pitch control also can be used to prevent the generator from being overpowered by stalling the blades. With passive stall, the blades are bolted in place. The blades are designed to stall at high wind speeds to prevent the generator from being overpowered. Most turbines today come with active pitch control.

- **Generator type:** Should the generator be synchronous or asynchronous? A synchronous generator, which is connected to the grid, will run at a fixed speed. If the torque going to the generator is increased, the magnetic forces of the generator will resist an increase in speed. As a result, a gust of wind will cause large stresses on the wind turbine's drive train. An asynchronous generator allows for a limited amount of slip or variation in generator revolutions per minute. Adjusting the resistance in the rotor windings can change the amount of slip. Therefore, the rotor revolutions per minute can increase a limited amount with a gust of wind, which reduces drive train wear and tear.

- **Pole-changing, two-speed generator:** If a grid-connected generator can change the number of poles it uses, it can change its synchronous revolutions per minute. Therefore, the turbine will have two generators in one. Of course pole changing increases the generator cost.

- **Speed:** Should the turbine be fixed speed or variable speed? If a turbine is connected directly to the grid, the rotor and generator must turn at a fixed speed to produce power at 60 Hz. If the turbine is connected to the grid indirectly, the rotor and generator can rotate at variable speeds. Allowing the rotor to rotate at various speeds results in a more efficient capture of wind energy and less stress on the turbine drive train due to wind gusts. However, the power from a variable-speed system must be rectified. Additional equipment is needed to increase the power, which results in additional cost, but it is not possible to achieve 100 percent efficiency.

  o **Fixed-rotor revolutions per minute (for fixed-speed wind turbine):** For a fixed-speed turbine, there will be an optimum rotor revolutions per minute that is dependent on the site's wind distribution. A large rotor-to-generator ratio captures more energy at low wind speeds, whereas a small rotor-to-generator ratio captures more energy at high wind speeds. The ratio must be optimized for site-specific wind speed distribution. Older-generation wind turbines had smaller capacities (that is, small rotor, large generator) and had less frequent failures. Newer designs, which are much larger than in the past, are designed with a smaller margin of safety (that is, larger rotors and similar-size generators). As expected, failure rates for gearboxes have increased because of this smaller margin of safety (see Chapter Four, "Gearboxes, Gear Ratios, and Failure Mechanisms").

- **Cradle-to-grave design concept:** This concept allows for flexibility in redesign because of problems in the field or in manufacturing processes, a flexibility that is possible because of the constant introduction of new technologies and review of current designs.

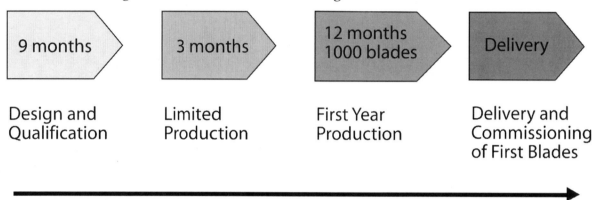

9 months — Design and Qualification

3 months — Limited Production

12 months / 1000 blades — First Year Production

Delivery — Delivery and Commissioning of First Blades

Total development cycle + 1000 blades = 2 years

- o **Engineering limitations for cradle-to-grave concept:** After 24 months, a new blade has been developed and 1000 have been delivered to sites and are flying. Suppose that a flaw is now reported in one of these potential problem areas:
  - Design
  - Structure
  - Materials
  - Manufacturing
  - Quality

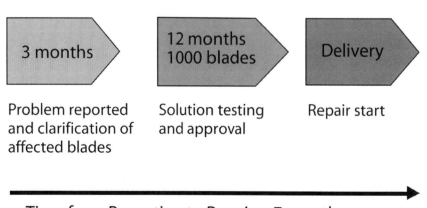

3 months — Problem reported and clarification of affected blades

12 months / 1000 blades — Solution testing and approval

Delivery — Repair start

Time from Reporting to Repair = 7 months

If this should occur, from first manufacturing to the beginning of repair has taken 19 months. During this time, potentially 1585 blades have been made and are now in need of repair. These could be (and generally are) distributed to remote sites, in a variety of countries, and thus infrastructure for repairs is limited.

# *Design Challenges*

### Challenges for Blade Producers

Currently, wind turbine blades are being manufactured at lengths as large as 300 ft in diameter, and there are prototypes in the range of 350 to 400 ft. As blades become larger, the need for more and lighter materials to compensate for the size will increase. New materials and manufacturing methods provide the opportunity to improve wind turbine efficiency by allowing for larger, stronger blades, but this presents certain challenges to turbine blade designers.

- **Blade weights and materials:** One of the most important goals when designing larger blade systems is to keep blade weight under control. The larger the blade, the heavier the demand on the service of the turbine, so it is beneficial to keep the weight in check as designs continue to grow. Current manufacturing methods for blades involve various proven fiberglass-composite fabrication techniques. Many different methods are available for blade manufacturing, from the hand-layup, open-mold, wet process to variations of that same process. Another available option is vacuum-assisted resin transfer molding. Ultimately, these options are nothing more than a variation on the same process.

  Epoxy-based composites are of greatest interest to wind turbine manufacturers because they deliver a key combination of environmental, production, and cost advantages over other resin systems. Epoxies also improve wind turbine blade composite manufacture by allowing for shorter cure cycles, increased durability, and improved surface finish.

  Carbon fiber–reinforced load-bearing spars have recently been identified as a cost-effective means for reducing weight and increasing stiffness. The use of carbon fibers can result in a reduction in total blade mass and a decrease in cost compared with a 100 percent fiberglass design. The use of carbon fibers has the added benefit of reducing the thickness of fiberglass-laminate sections. Wind turbine applications of carbon fiber may also benefit from the general trend of increasing use and decreasing cost of carbon-fiber materials.

- **Large differences between "qualification" blades and actual production blades:** The qualification blade may be more detailed or have more differences in its design, as opposed to a production blade that would be manufactured by a template or jig. The qualification blade will most likely be manufactured by hand and become the mold. The inherent challenge in this design issue is that the original, because it is handmade, could have slight variations from the mass-produced blades that will follow.

- **Large differences between qualities of materials used with different manufacturers:** Material quality may fluctuate from manufacturer to manufacturer.

This variability in quality can affect the efficiency of one blade over another, even though the initial design is the same for both blades.

- **Composites are not metals:** Composites are much more complex to deal with than metals. Composites have variations within the matrix of the composites themselves. For example, composites absorb water, which could create swelling in humid environments. The coefficient for thermal expansion in composites is also much higher than in metals. Here are some considerations for composites:

  ○ The materials set within their molds as the structure is built, leaving little time for adjustments.

  ○ Globally, a low skill base on composites exists and there is difficulty in transferring skills from other industries.

  ○ Composite materials are generally custom-made, making quality difficult to determine.

  ○ Composites do not perform well in anything but well-controlled manufacturing environments. Factories manufacturing blades are appearing across the world in "suitable" sites, yet the generally unskilled workforce with no knowledge of composites and limited training creates inherent problems. Global dispersion and required "local" manufacturing lead to products that vary by site, which leads to confusion when problems occur.

  ○ There is a high demand for blade output from the day of start-up yet a slow realization of (and response to) problems that arise because of the limited amount of specialists available to help.

  ○ More blade failures are expected in the field over the coming years: Turbine buyers and owners should apply pressure to their suppliers to provide the optimum cost and reliability mixture—not just the cheapest blade option. Manufacturers must know their blades and understand how they are made, as well as be able to take advice on manufacturing methods, materials, and engineering. Energy companies should demand quality and consistency of manufactured blades, and manufacturers must ensure that blades are thoroughly inspected before they leave factories, as well as before and after shipment and before installation. Additionally, the industry must accept that mature blade solutions take time to develop but are worth paying extra for.

### Aerodynamics

Engineers who design rotor blades are concerned with aerodynamic drag, which is always a consideration when an object moves rapidly through the air. Wind turbine rotor blades must have high tip speeds to work efficiently. Therefore, it is critical that rotor blades have low aerodynamic drag.

### Drag

The arrows in the following illustration of drag show the direction of the drag force, when the cross section of the rotor blade is moving toward the left. You can see that feathering the rotor, leaning it downward, increases wind speed, and thus lift is created on the bottom of the rotor, much the same as would be the case with an airplane wing.

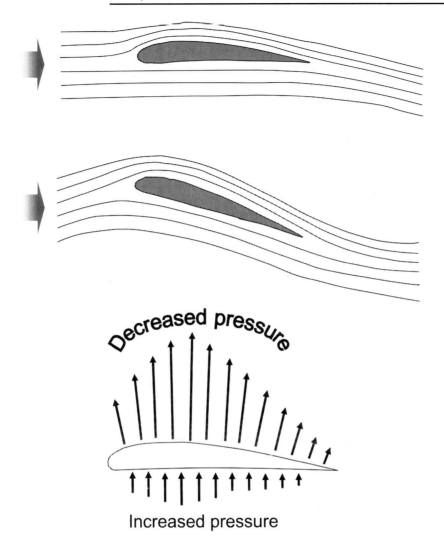

Decreased pressure

Increased pressure

- **Drag increases with the area facing the wind:** Drag is proportional to the cross-sectional area of the object in the wind. The higher the cross section, the higher the drag. Imagine holding your hand outside the window of a moving car. As you increase the angle of your hand to increase the cross section of your hand facing into the wind flow, the force required to hold your hand in place also increases. As you flatten your hand, the amount of the force required to hold your hand stationary decreases. Try it! Objects that have to move quickly, but with low energy use, through fluids should be designed so that they have small cross sections facing the current. For this reason, submarines are designed using elongated drop shapes.

- **Drag depends on the shape of the object:** Aerodynamics can be measured by a dimensionless constant, $C_D$. The size of the drag for a given shape is usually measured by the drag coefficient, $C_D$, which is defined as the drag force per area of a cross section of the object.

The two-part picture above shows one piece of wood from two different views. The frontal cross section for the first picture is very low and thus has a very small $C_D$. The second picture shows the piece of wood rotated 180 degrees and has a much higher frontal cross section. Such a shape has a very high $C_D$. An airfoil shape used on rotor blades typically has an extremely small $C_D$.

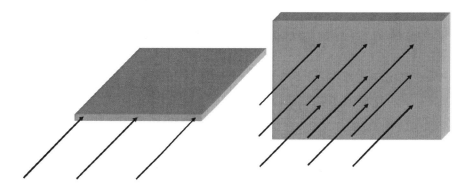

- Drag increases with the square of the wind speed: Both lift and drag force increase with the square of the wind speed, as shown by the following equation:

$$F_D = \frac{1}{2} \times \rho \times v^2 \times A \times C_D$$

$$= \frac{1}{2} \times 1.165 \, \text{kg}\!\big/\!_{\text{m}^3} \times \left(11.17 \, \text{m}\!\big/\!_{\text{s}}\right) \times 0.6567 \, \text{m}^2 \times 0.82$$

$$= 39.14 \, \frac{\text{kg} \cdot \text{m}}{\text{s}^2}$$

$$= 39.14 \, \text{N}$$

$$F_D = \frac{1}{2} \times \rho \times v^2 \times A \times C_D$$

$$= \frac{1}{2} \times 1.165 \, \text{kg}\!\big/\!_{\text{m}^3} \times \left(11.17 \, \text{m}\!\big/\!_{\text{s}}\right) \times 0.3716 \, \text{m}^2 \times 1.05$$

$$= 22.15 \, \frac{\text{kg} \cdot \text{m}}{\text{s}^2}$$

$$= 28.36 \, \text{N}$$

where:

$F_D$ = drag force (N)

$\rho$ = air density (kg/m³)

$v$ = velocity (m/s)

$A$ = frontal cross section (m²)

$C_D$ = drag coefficient (N/m²)

In more practical terms, a car with a 150-hp engine will use about 6 hp to overcome air drag and about 14 hp for mechanical propulsion when it is being driven at 50 mph. If that same car increases its speed to 130 mph, it will use 112 hp to overcome air drag and 40 hp to overcome rolling resistance. Additionally, drag will increase if air density or frontal cross section increases. The formula for the drag force is fairly logical when you compare it with the formula for the power of the wind. The variable $v^2$ is used in the formula because we are dealing with a force. If we had looked at the power loss from drag instead, we would have multiplied by $v^3$.

- **High lift-to-drag ratio needed on airfoils for rotor blades:** The rotors on modern wind turbines have very high tip speeds for the rotor blades, usually around 75 m/s (270 km/h, or 164 mph). To obtain high efficiency, it is therefore essential to use airfoil-shaped rotor blades with a very high lift-to-drag ratio—that is, rotor blades that provide a lot of lift with as little drag as possible. This is particularly necessary in the section of the blade near the tip, where the speed relative to the air is much higher than it is close to the center of the rotor. For wind turbines with a low tip speed, it is not necessary to use top-quality airfoils.

- **Drag increases with the density of air:** Referring back to the drag equation, note that both lift and drag increase in proportion to the density of air. Cold air thus produces more drag than hot air does. Air density, then, is proportional to temperature. Refer to the table below to see the difference in air density as temperature changes.

- **Drag coefficient varies with the roughness of the object surface:** Just as for lift, drag may vary quite dramatically with the surface roughness of the object. Normally it is desirable to have smooth, clean surfaces to minimize drag. The smoother the surface, the lower the amount of drag, which ordinarily increases lift.

| Density of Air at Standard Atmospheric Pressure | | |
|---|---|---|
| Temperature, Celsius | Temperature, Fahrenheit | Density (i.e., Mass of Dry Air) in kg/m³ |
| -25 | -13 | 1423 |
| -20 | -4 | 1395 |
| -15 | 5 | 1368 |
| -10 | 14 | 1342 |
| -5 | 23 | 1317 |
| 0 | 32 | 1292 |
| 5 | 41 | 1269 |
| 10 | 50 | 1247 |
| 15 | 59 | 1225* |
| 20 | 68 | 1204 |
| 25 | 77 | 1184 |
| 30 | 86 | 1165 |
| 35 | 95 | 1146 |
| 40 | 104 | 1127 |

*The air density of dry air at standard atmospheric pressure at sea level at 15°C is used as a standard in the wind industry.

- **Drag depends on the Reynolds number:** In reality, there is not just one but two kinds of drag: pressure drag and friction drag. At very low speeds, and for small objects—say, dust particles—the friction drag dominates. At high speeds and/or large object sizes, pressure differences dominate. The drag coefficient for an object will therefore depend on which type of flow is dominating. A microscopic parachute will not work like a large parachute. Fortunately we are able to predict which type of flow we are dealing with if we know the so-called Reynolds number for the experiment. The Reynolds number is defined as

$$\text{Re} = \frac{v \times \text{L}}{\left(\dfrac{\mu}{\rho}\right)}$$

Where:

Re = the Reynolds number, which is dimensionless, meaning that it is a ratio of two quantities with the same unit

$v$ = the relative velocity of the fluid in meters per second

L = the characteristic length, in this case the largest cross section of the frontal area in meters

μ = The viscosity of the fluid in N*s/m². The viscosity of air, also called the dynamic viscosity of air, is 1.8¥ 10⁻ ⁵ at 15°C and atmospheric pressure at sea level

ρ = The density of the fluid in $kg/m^5$

The value in the denominator of the fraction (μ/ρ) is called the kinematic viscosity of air. When the kinematic viscosity is high, the laminar (or smooth) flows dominate.

Thus, if the Reynolds number is very small, less than 1, you can ignore pressure drag and concentrate on friction drag. If the number is large, more than 100, you can ignore the friction drag and look at pressure drag only. When close to the surface of the object, friction drag and viscosity are always important.

Now that you have an understanding of how drag affects wind turbine efficiency, look at the following problem using the concepts you have learned.

### Sample Problem

Airflow across blades can have dramatic impact, depending on geometry. Determine the initial drag force of the two shapes shown using the given information and an air temperature of 86°F. Which shape has less drag, and why? Determine the drag force if the wind velocity is increased by 65%. What differences are noted?

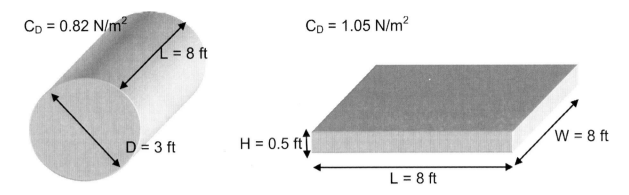

Given:

Initial velocity = 25 mph
1 mile = 1.609 km
1 ft$^2$ = 0.0929 m$^2$

$$T_C = \left(\frac{5}{9}\right)(T_F - 32)$$

## *Materials Used in Wind Turbine Manufacturing*

It seems as if every new day brings a larger turbine design. There are significant challenges with designing and building these larger-capacity turbines. One such challenge is larger blades to drive the larger turbines. Covering a larger sweep area effectively increases the tip-to-speed ratio, assuming a constant wind speed, which ultimately increases the ability of the turbine to generate more electricity.

As you have seen, aerodynamics is important in the design of wind turbines. Aerodynamics is affected not only by shape but also by materials. Therefore, research and development on the best materials for wind turbine use is an ever-growing field. Currently, there are several types of materials used in a wind turbine, with the lightest being used in the manufacture of blades. The most common types of materials are steel, concrete, and composites.

- **Steel:** An alloy of iron that contains 0.8% to 1.5% carbon, steel can be worked in either a heated or cooled state. The properties of steel can be changed by altering the carbon content and heat-treating the material. Steel is one of the most widely used materials in wind turbines. It is used for many structural components, such as the tower, hub, main frame (bedplate), shafts, gears, and gearbox cases; even the rebar in the concrete is steel. Some older nacelles are made of steel.

- **Concrete:** The tower foundation is a concrete, steel-reinforced structure to which the tower base is attached. It serves as the anchor for the wind turbine and is generally composed of two major designs for land-based turbines. One is the spread-foot design; the other is the can design.

- **Composites:** Composites are the primary material used in blade construction and in many nacelles. Composites are materials composed of at least two dissimilar materials. Most commonly in the wind industry, that means fibers held in place by a binder matrix. A binder matrix is something (such as tar or cement) that produces or promotes cohesion in loosely assembled substances. Resin is the binder in a blade. Blades are composed of glass fibers, which are formed by spinning glass into long threads. The fibers are usually combined in forms commonly referred to as molds. Then a resin is applied to create a fiberglass composite. Three common resins are currently used in fiberglass composites: unsaturated polyesters, epoxies, and vinyl esters. Polyesters have been most common in the wind industry because they cure relatively quickly and have a low cost. Epoxies are stronger and have lower shrinkage on curing, but they cost twice as much as polyester. Vinyl esters are epoxy-based resins, and they have similar properties to epoxies and are somewhat lower in cost and have a shorter cure time than epoxies. They tend to be very stable and are widely used in marine environments.

## Common Blade Problems

On the job, one of your important duties will be to inspect the turbine blades for damage. When blades are not turning, or are not turning efficiently, energy is not being produced, costing the tower owner money. You must be familiar with some of the common issues encountered in blade design as well as the ways in which to inspect blades.

Some of the most common blade issues you may encounter are

- **Manufacturing issues:** As you saw in the section on blade design in this chapter, several issues may occur at a blade's origination. A mold may have warped because overuse, or a specific type of blade may have functioned well as a prototype but problems could surface during mass production.

- **Transportation and construction:** Blades are built far from the sites where they will be installed, then shipped by truck to the construction site. In some cases, intermediate products are created in the United States and shipped back overseas for blade production. For example, in Abilene, Texas, Zoltek produces carbon-fiber thread, which is subsequently shipped to Germany for blade production, and the blades are in turn shipped back to Texas by boat, then trucked to the construction wind turbine site. With so much moving around, the blade or blade materials could be damaged.

- **Lightning damage:** Lightning strikes are a wind turbine's worst enemy, as you may recall from Chapter Two, "Safety on the Job." Without effective lightning protection, both the blades and the turbine itself can be severely damaged by the powerful energy surges in lightning.

  A lightning strike on an unprotected blade can lead to temperature increases of up to 30,000°C and result in an explosive expansion of the air within the blade. This can cause damage to the blade surface, delaminating, cracking

*Lightning damage*

on both the leading and trailing edge, and melted glue. Lightning strikes can also cause hidden damage that over time will result in a significant reduction of the blade's service life.

Wind turbines are particularly complicated to protect because they have so many different components—including nonconducting composite materials such as glass-reinforced plastic. Any lightning protection system must therefore be sufficiently comprehensive to take into account all of the parts. With the introduction of larger and taller turbines, lightning strikes to wind turbines have become more prevalent. Massive blades often have a receptor at the tip, which can channel the lightning into the proper wires and onward to the ground.

- **Leading-edge erosion:** Leading-edge erosion, as seen in the photo below, can be caused by a number of things, including wind, flying debris, and ice. As the blade erodes, the surface area is changed, allowing for less-efficient aerodynamics, as well as the possibility of delamination.

*Leading-edge erosion*

- **Debonding at spar or trailing on leading edges:** Often caused by extreme ambient heat but more likely by a preexisting flaw, this is a process in which materials that are lapped over each other begin to come apart.

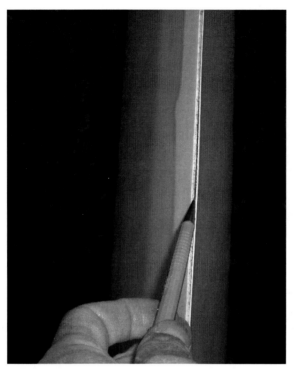

*Debonding at spar*

*Cracked trailing edge*

## *Inspections Detect Blade Problems*

As you can see, the blades are not only a primary structure in a wind turbine; they are also the structures with which many problems are likely to arise. Proper blade inspection is key to keeping a turbine online and operational. When do inspections occur?

- **Before installation to detect manufacturing, shipping, or construction issues:** Many blade problems are not found until after installation, when operation issues show up. It is better to inspect blades on the ground, when it can be easily accomplished. At this point, a technician is looking for manufacturing defects inside and out.
- **Near the end of the warranty period:** It is crucial to inspect a blade as the end of its warranty approaches. The condition must be verified, any damage reported, and warranty repairs completed before the hand-off. Nonwarranty repairs can be completed at the time of this inspection.
- **Periodically during service years using various forms of nondestructive test methods:** See the next section for details.

## *Inspection Techniques*

Inspection techniques range from using ultrasound and infrared light to conducting visual inspections. Some of the inspection techniques you will encounter are

- **Visual inspection:** In this technique, you depend on your eyes, binoculars, and/or high-resolution cameras to detect visible flaws in a blade.
- **Tap test and probing:** You should tap the blades before installation, listening for changes in sound that might indicate a structural weakness, such as an air bubble that could lead to debonding later on.
- **Thermographic camera:** When you use a thermographic camera to look for stresses in the laminations, you will see that stress creates heat, which shows up on the thermographic image, as in the illustration below.

- **Ultrasound:** Ultrasound shoots sound waves through the blade and measures the bounce as the wave returns. If an imperfection exists, the ultrasound detects it because the blade density is different and the sound wave therefore reacts differently in that area.

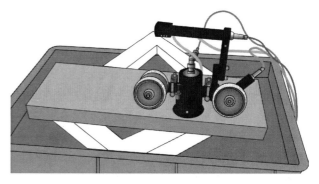

- **Infrared imaging of interior:** With this technique, infrared light is used to detect imperfections.

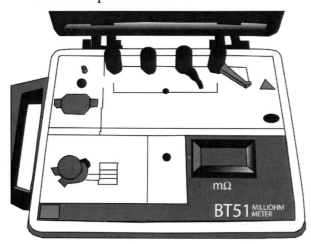

- **Lightning system continuity check:** In this inspection technique, you are looking for a break in the circuitry or a discontinuity between electrical points A and B. Each blade has a receptor on it that must be tested.

- **Inspection documents and database:** Material history is an important tool in diagnosing problems, or potential problems, on a wind turbine. You should always check your findings against the material history, as was stated in Chapter Four, "Gearboxes, Gear Ratios, and Failure Mechanisms." A savvy technician may realize that a trend indicating a potential problem is developing and be able to take appropriate preventive measures, saving time and money for the tower owner.

## Specialized Repair

Each blade on a turbine is unique, so the repair process must be determined for each blade, by the damage present and the type of material used for construction. Repairs begin with selecting a repair material, often a bonding agent that can be used to fill a damaged area. Next, the damaged area must be prepared for lamination or bonding. You will have to sand the area to make sure there are no protruding fibers, then clean the sanded area. After sanding, apply the proper laminates, making sure that fibers are aligned properly. Then shape the blade to restore it to optimum aerodynamic shape by sanding and feathering the filler material. Finally, apply a sealing coat of paint to protect the repaired area against future damage.

## *Causes of New Wind Blade Problems*

Being fresh from the manufacturer does not mean that a turbine blade will not have problems. New blades can have problems because of any of the following:

- **Speed to market:** This is a business term that means "accelerated production due to market demand." This can result in a variety of problems, from lack of quality control to blade molds that warp because they are not allowed them enough time to cool down between uses.

- **Lack of proper testing time:** Shortened time demands can create an environment in which there is no time for adequate field and fatigue testing.

- **Cost factors:** Pressure to design lower-cost blades may mean a compromise in quality.

- **Safety issues:** Speed to market may result in a reduction of safety margins.

- **Scaling up is not easy:** What works in prototype on a small scale may not work on an actual scale. Only adequate time for testing can prove the success of a new design.

## *Failure Mechanisms in Wind Turbines*

When a wind turbine suffers downtime, the owner or operator loses money, so the pressure is on the wind technician to get the tower back online as quickly as possible. You must be aware of the types of failures that can occur in a wind turbine and have the knowledge and skills not only to identify the failure but also to repair the turbine as quickly as possible. In this section, you will learn more about the two primary failure mechanisms you may encounter on the job: electrical component and mechanical component heat-related failures.

### Electrical Component Heat-Related Failures

Electrical component failure, or heat damage related to an electrical failure, is a primary cause of wind tower incapacitation. There are various common reasons for electrical component heat failures:

- **Wrong conductor sizing:** The NEC (National Electrical Code) provides minimum size requirements for conductors to prevent overheating and fire. Insulation type, ambient temperature, and conductor bundling are three primary factors in determining how big a conductor has to be for it to safely carry the current imposed on it.

  A key concept in conductor sizing is ampacity. The ampacity of a conductor is the amount of current the conductor can carry continuously under specific conditions of use (see "References and Additional Resources" at the end of this chapter for links). The ampacity of a conductor is not what size breaker can be used to protect the wire; it is simply the amount of current the conductor can carry. It is important to recognize this subtlety.

When conductors are bundled together, they lose some of their ability to dissipate heat. In the NEC, the allowable ampacity starts dropping when four or more current-carrying conductors are bundled together for more than 24 inches.

- **Poor termination:** A poor electrical connection under load will increase in temperature. Arcing occurs, and eventually the connection can fail or, at worst, cause a fire, resulting in an expensive shutdown of the associated equipment or machinery.

- **Component heat production:** Electrical equipment generates large amounts of heat, often referred to as $I^2R$ losses (current [2] • resistance). $I^2R$ losses can be defined as the energy generated or lost as heat because of internal resistance.

- **Current moving through copper wires:** This movement can generate large amounts of heat. With regard to wind turbines, there are numerous cables and equipment that, along with electrical generation, give rise to plenty of opportunity to generate heat within an enclosed space.

- **Lack of cooling:** Remember that the heat generated by electrical current is often referred to as $I^2R$ losses. $I^2R$ losses can be defined as the energy generated or lost as heat because of internal resistance. If equipment is not effectively cooled, the heat generated can lead to component failure, both directly and indirectly as a result of heat. Additionally, cooling can be addressed at the component scale or at the turbine itself. A direct correlation would be a computer that has its own heat sink and fans to cool the microprocessor. The computer tower itself will remove heat from inside the tower to the room it is located in. If the room does not have adequate cooling or ventilation, it is only a matter of time before the heat generation will lead to component failure.

- **Poor cabinet ventilation:** As in the example of the computer, cabinet ventilation is directly related to cooling electrical components. If there is adequate cooling but no path for the heat to be removed via ventilation, the same component failures can occur. Lack of cooling and poor cabinet ventilation often go hand in hand when it comes to heat-related failures.

- **Poor system design:** Engineers spend an extensive amount of time designing systems that will provide adequate cooling and proper ventilation paths. However, there can be factors related to schedule, cost, poor analysis techniques, improper testing, and so on that can lead to improper design of components. Additionally, if the system is not properly specified during conceptual and preliminary design phases, there can be the potential for an improperly designed system, which leads to heat-related failures.

- **Extreme ambient conditions:** All systems, when initially designed, are structured for a certain set of working conditions (heat, humidity, dust, ice, etc.). Most applications will try to design for ambient conditions to the extent possible. However, there may be limitations within the design specification that can restrict the components' ability to work in standard ambient conditions. An example would be if it became cost prohibitive to include a certain fan that could move a greater amount of air. As a result, the engineer might specify a different fan that be more readily available or affordable but moves significantly less air.

Even if the system is properly specified during design, there may be instances in which those design conditions are exceeded. In most cases, a safety factor is calculated into a design, but that invariably increases the cost of the equipment.

### Mechanical Component Heat-Related Failures

Mechanical components, like electrical components, can fail for a number of reasons. Technicians should remember to always check the material history data on a given tower to see if a trend has been developing. These are some of the common mechanical component failure mechanisms:

- **Poor material design (heat-treating protections):** It is important to accurately identify the proper material when designing components. Not only should the material be able to effectively remove heat as specified but also it must meet cost, availability, and manufacturing goals. Some materials have excellent heat-removal qualities but may be very expensive to machine and/or manufacture. Titanium is an example; it has high strength and excellent corrosion-resistance properties but is cost prohibitive and difficult to weld and machine.

- **Excessive operation:** Mechanical components are designed for operation, but as with all materials, they have limitations, and if not properly treated, they will cause premature failure. Typically, excessive operation can be directly attributed to lack of lubrication. However, if you assume that there is adequate lubrication, the component itself can still generate large amounts of heat. As an example, your car engine has many ways of cooling itself, from the oil system to the radiator, yet its normal operating temperature still is considered hot to human touch. Most metals lose their strength as their temperature increases, which leads to potential fatigue over large cyclic stresses. Imagine a coat hanger being bent over and over; eventually, the wire heats up and becomes ductile to the point of failure.

- **Lack of lubrication:** This is probably the most common occurrence of failure within mechanical systems. There are many issues that can cause failure, including

  ○ **Contaminants:** Any foreign matter residing within the lubricant has detrimental effects on the lubricating properties. They can be introduced because of the operating environment (for example, West Texas dust) or can be inherent to the system as components deteriorate over time (for example, gears chipping).

  ○ **Water:** Some systems use water as a heat sink medium to remove heat from oil. In some cases, there can be a leak, allowing water to be intro duced into the system, so that the oil loses its lubricating properties.

  ○ **Poor filtration:** This correlates to contaminants in the system. If the filtration medium breaks down or is no longer effective, the oil will not per form its desired function.

○ **Poor maintenance of lubricants:** Oil is not a permanent medium and has to be periodically changed, as dictated by maintenance schedules. If the schedule is not adhered to, the oil will eventually break down and not perform its intended function, leading to premature failure of a component. The primary purpose of lubricants is to remove heat and reduce the friction between two moving components. The two moving components can be rotating equipment (a shaft rotating within a bearing or a hydraulic pump for the power plant).

○ **Lack of cooling systems:** Cooling systems work hand in hand with lubrication systems. Typically, water is used as a medium to dissipate the heat that oil removes from any piece of equipment that generates heat, whether by friction, electricity, or other means. Each system is simple in design, yet there are many actions that can lead to premature failure due to overheating of components. Other types of cooling systems, such as a radiator system or cooling fan in an up-tower controller, use air as a medium to remove heat.

○ **Extreme ambient conditions:** As stated earlier, extreme ambient conditions such as dirt, heat, cold, and wind can contribute to poor performance of mechanical systems. Extreme temperature has the ability to expand and contract seals, a situation that can introduce contaminants or cause leaks. High winds can create differential pressures that can lead to contamination of a system.

## Conclusion

Effective wind turbine operation begins with excellent design and manufacturing. Lightweight materials must be used, and engineers have to think outside of the proverbial box to find new and better solutions and sustainable designs. Still, costs in design and manufacturing are a determining factor and must be considered. Because of cost, manufacturing, and/or design limitations, combined with the failure mechanisms inherent in any electromechanical equipment, wind turbine technicians will have job growth and stability in the long term. Technicians who have learned the material in this book will have a strong foundation for a long, successful career.

## References and Additional Resources

### National Electrical Code Information
http://en.wikipedia.org/wiki/National_Electrical_Code_(US)

http://www.codebookcity.com/codearticles/nec/necarticle100.

### Wind Turbine Blade Material Information
http://pepei.pennnet.com/display_article/282536/6/ARTCL/none/none/1/New-Structural-Materials-for-Wind-Turbine-Blades/

http://composite.about.com/od/inthenews/l/blsandial1.htm

http://www.lmglasfiber.com/dalmg/composite-wind-turbine-blades.htm

**Blade Failure Information**

http://www.sciencedirect.com/science?_ob = ArticleURL&_udi = B6V2X-4RY8SW7-3&_user = 10&_rdoc = 1&_fmt = &_orig = search&_sort = d&view = c&_acct = C000050221&_version = 1&_urlVersion = 0&_userid = 10&md5 = 7b918d1e5145a850a59d342a203987a7

http://www.nawindpower.com/page.php?5

Sorensen, Brent. Designing against different failure mechanisms in wind turbines. Availablefromhttp://www.extra.ivf.se/eccm13_programme/abstracts/Plenary%20Bent_Sörensen.pdf. Accessed May 25, 2009.

**Wind Turbine Blade Inspection**

Mattei, Christophe, and Dahrèn, Magnus. n.d. Advanced ultrasonic NDT techniques for inspection of wind turbine blades. Linköping, Sweden: Bodycote Materials Testing AB. Available from http://www.bodycote-mt.se/bodycote_eng/pdf/pdf_ladda_ner/eng/Bodycote_ENG_011.pdf. Accessed July 24, 2009.

http://www.dantecdynamics.com/Default.aspx?ID = 838&M = News&PID = 4492&NEWSID = 410

http://earthblips.dailyradar.com/story/riwea_the_rope_climbing_wind_turbine_inspection_robot

## *Review Exercises*

1. How tall is the tallest wind turbine in the world, and how many homes can it generate power for?

2. List and describe three important design variables to be considered for wind turbine blades.

3. Explain the cradle-to-grave design concept. How does this concept affect currently online turbines?

4. What is the material most commonly used in blade manufacturing? Why is it the most common?

5. Describe the potential differences between a qualification blade and a production blade.

6. List the common blade problems you are likely to encounter and write a brief description of each one.

7. Why should blades be inspected near the end of a warranty period?

8. List three techniques used in blade inspection and describe how they work.

9. Why is it crucial to review the material history data sheets for each tower you inspect and update them with current information?

10. Describe the process of repairing a damaged blade.

11. List and explain three causes of electrical failure.

12. List and explain three causes of mechanical failure.

## *Application Exercises*

1. Using the Web sites whose addresses are listed in the "References Additional Resources" section at the end of this chapter, as well as at least three Web sites you access on your own, research wind turbine blade failure mechanisms and write an essay about the most common failures and what is being done in the industry to combat the problem. Cite your sources.

2. Using the same geometries from the sample problem on drag, determine the Reynolds number and what, if any, impact it has on friction and pressure drag. Do the geometries play any role in determining the Reynolds number?

   Given:
   Initial wind velocity = 11.17 m/s
   Final wind velocity = 18.43 m/s
   Length of cylinder = 8 ft
   Length of flat plate = 8 ft
   Kinematic viscosity = $1.8 \times 10^{-5}$ N s/m$^2$
   Air density = 1.165 kg/m$^3$

3. The final drag force acting on a cylinder is 13.5 lbf. Knowing the diameter of the cylinder is 2.5 ft and the drag coefficient is 0.82, determine the final wind velocity. What is the initial wind velocity if the final drag force was 30% more than the initial drag force? Assume that air density is at a standard temperature of 15°C.

   Given:
   $F_{D \ Final}$ = 13.5 lbf
   $C_D$ = 0.82 N/m$^2$
   $D$ = 2.5 ft
   $\rho$ = 1.225 kg/m$^3$
   1 lbf = 4.448 N
   1 m = 3.28 ft

# Index

# About the Authors

*Keith Plantier* is the Program Director for the Texas Wind Energy Institute at Texas State Technical College (TSTC). Keith began his career in the US Navy as a nuclear mechanic and received an Associates of Science in Nuclear Technology from Thomas Edison State College. Keith left the Navy after nine years of service to continue his education at Texas Tech University and received both his Bachelor and Master's degrees in Mechanical Engineering. After graduation, Keith became a design/project engineer at Northrop Grumman working with manned/unmanned undersea systems. Keith returned to West Texas to become Program  Director for the Texas Wind Energy Institute at Texas State Technical College. Keith has over nineteen years experience in power plant operation, maintenance, and design.

 *Karen Mitchell Smith* is a graduate of Texas Tech University. She has been an English and Spanish teacher, a motivational speaker on the importance of post-secondary education, and a technical college recruiter. An award-winning writer and editor for more than 20 years who has an abiding concern for students and a wish to see them succeed, she writes on many education-related topics. Karen's website is at www.topshelfediting.com, and she can be reached for questions or comments at karen@topshelf-editing.com.

# About TSTC Publishing

Established in 2004, TSTC Publishing is a provider of high-end technical instructional materials and related information to institutions of higher education and private industry. "High end" refers simultaneously to the information delivered, the various delivery formats of that information, and the marketing of materials produced. More information about the products and services offered by TSTC Publishing may be found at its Web site: http://publishing.tstc.edu/.

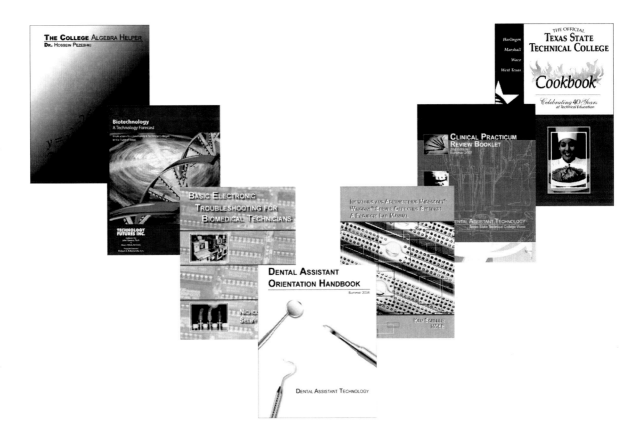